# The Tale of the Tailless Sperm

Joanna Infeld

Copyright © 2020 by Joanna Infeld

All rights reserved under International Copyright Conventions. No part of this book may be used or reproduced in any manner whatsoever without written permission from the publisher.

Printed in the United States of America

Kora Press ® is a federally registered trademark.

Kora is a meditative journey around
a mountain, stupa or sacred site.

Published by Kora Press ®
*www.KoraPress.com*

# Contents:

Chapter One: Jenny Can't Get What She Wants..............5

Chapter Two: Dick Hears the Truth................17

Chapter Three: Where Have All the Tails Gone?...........23

Chapter Four: The Tale of Two Non-Existent Tails.......29

Chapter Five: The Race Toward Success........................45

Chapter Six: A Disturbing Trend......................89

Chapter Seven: The Plot Thickens....................97

Chapter Eight: The Search Is On....................109

Chapter Nine: Jenny's Dilemma......................127

Chapter Ten: A Global Issue............................131

Chapter Eleven: Epilog....................137

# CHAPTER ONE

## Jenny Can't Get What She Wants

Dick rolled over and smiled. There was something delicious and naughty about making love in the morning, before work. It seemed to set him up for the whole day, and although slightly tired, he also felt the tensions brought about by the thought of another day at the office slowly oozing out of him, as if he were on vacation, or it was the weekend, or as if he didn't really care. Jenny had gone to the bathroom, come back, snuggled under the cover for another five minutes of warmth before the alarm clock on their bedside table would ring in that horrible noisy tone—the only one that could ever get her out of bed on time. She had turned her back to him and he watched now the curve of her back, as it moved slightly with the rhythm of her breathing. "Pretty Jenny" he always called her, for she really was rather beautiful. The shiny chestnut hair fell on the pillow beside him and he could smell it now, that familiar scent of the shampoo she used.

"Jenny, it's time to get up. It's time to go to school." He always used to say that, though she didn't really go to school. She taught four and five year olds and it was really pre-school. She just groaned in reply, as if requesting for him to leave her alone, and he reached to the bedside table to turn the alarm clock off, as it was his responsibility to get them both up in time for work. So as usual, he was the first one to throw back the covers and sit upright with his feet on the floor, stretching his arms in front of him as he did so. Then he was up, slowly adopting his businesslike posture, running the shower, getting dressed and with each item of clothing he put on, he also put on in increments a different look—the polished look of a salesman who was professionally confident, successful and effective.

"Come on, Jenny," he said. "It really is time to get up!"

She turned around and smiled at him.

"You go and put the coffee on and I'll be right there," she said, propping herself up on her elbow.

"All right, but I won't wait for you." He always said that and he always did wait. He hated last minute rushes.

As they went out to the car, Jenny held a piece of bread and jam in one hand and her make up bag in the other, with her purse slung over her shoulder. It was like this every day—she never had the time to do all the things she wanted to do before getting out of the door. Dick threw his briefcase on the back seat and sat behind the wheel.

"Have you got everything?" he asked. This, too, was part of the morning ritual.

"Yes, go. Move it," she said with her mouth full. And so he started the ignition and slowly backed out of the driveway, ready for another day.

# Jenny Can't Get What She Wants

"Dick, can you come into the boardroom?" his boss's voice said over the office intercom. "I want to show you something." He sensed the note of urgency in Mr. Simpson's voice.

"Coming right away, Sir."

He walked down the hall and as he did so, he saw the other salesmen pouring out of their offices, all with a note of urgency too, correcting their ties and smoothing down their hair as they went. The two salesgirls also appeared, both wearing suits and chatting as they walked together.

"This must be important," he thought just before he was about to enter the boardroom. He was joined by the two guys from the Advertising Department. They were carrying large portfolios and it felt like something new was going to be announced.

In the boardroom Mr. Simpson was sitting at the head of the table, as he always did, with a morning coffee in his white bone china cup and matching saucer, with his secretary by his side. They looked as if they had been there for hours. Perhaps they had; it would not be the first time. The salesmen and ladies, the sales manager and deputy manager, the advertising guys were all huddled around the coffee machine set up on the sideboard by the canteen staff at the far end of the room. They were pouring themselves mugs of coffee and exchanging greetings. They then began filling out the seats around the large oblong boardroom table. The coffee machine was one of the advantages of meeting in the boardroom, and once again, Dick was impressed how well this place was run. Not like Mac's Timber Warehouse where he first started out and

# The Tale of the Tailless Sperm

where everybody did everything or you were soon out of a job.

As Dick took his place at the table, a quiet descended and Mr. Simpson began. "We have a new line to show you," he said, "and this one will be a hit." He looked at the new advertising genius who was sitting on his left and had been "stolen" from a fashionable downtown advertising agency just three months before.

Andy stood up. This was his first important presentation, his first star performance. If he could make this one stick, he would become one of the guys, a team player, and perhaps even a future director or partner. He straightened his rather large tie as he stood up.

"Ladies and gentlemen," he said, "I am proud to present our new advertising campaign. Zigzag is entering the New Age with a bang and it will be taken up by all three of our factories—buttons in Hong Kong, haberdashery in Singapore and accessories in Manila. We are going to produce a new line of astrological buttons and accessories. Here are some samples." He placed his portfolio on the table in front of him, opened it up and pulled out a series of drawings in color. "Could someone please close the curtains?" he asked as he switched on the overhead projector and the enlarged illustrations appeared on the white wall behind him, depicting a range of buttons in many colors, large and small, all with astrological motifs. Then there were imaginative designs of ribbons, trimmings, edgings and laces, all with astrological signs and symbols. Andy was looking very pleased with himself, as if he had just justified his large salary; he was enjoying the exclamations stating intelligently, "Ooh," "That's neat," and, "People will love that."

"The thing about astrology," Andy explained, as each drawing appeared in turn on the overhead projector, "is that everyone has a sign and everyone knows what it is. Who doesn't read the predictions for their sign in the morning paper, even if they think it is nonsense?" He paused and looked around. No one responded. He then continued, "I thought so. People are very curious about what it all means and we are going to capitalize upon this fact. We will also produce explanation leaflets that will go with each collection. Perhaps we will also produce a regular newsletter and an optional astrological chart upon request."

Mr. Simpson spoke next. "We have produced leaflets and brochures showing the ranges and selections; we have price lists including our preferred clients' discounts and long term payment plans—all ready to go. So I want you"—here he turned to the salesmen and sales ladies—"to contact all our existing clients and the new accounts, as well as those we haven't dealt with yet, like New Age stores, speciality stores, craft markets, dressmakers, fabric stores and designers. In other words, get out there and sell. Next month we are having a two page spread in *Vogue* magazine and we want our astrology line to become the desire of every woman who wishes to be fashionable, every teenager who wants to hang out with the in-crowd, and every New Age dieting, meditating, regressing, spiritually aware middle-aged snob who wants to be seen wearing the right thing. This is the answer, and we have two contracts with popular news anchors and a pop singer to wear our astrological buttons for a month. That should be enough to get them noticed."

"Oh, how exciting," thought Dick, stifling a mental yawn. "I wonder who dreamed this one up." But, faithful to his career and vision of the future, he joined in the applause. He received his package, his quota of brochures, price lists, display photographs and samples, and he was now well equipped and ready to go.

Back at the office there was a message from Jenny waiting for him. He looked at his watch. She would be in class now, so there was no point phoning her. She sounded urgent. "Oh well, I'm sure it can wait." The most urgent requirement now was a coffee. He buzzed his secretary.

"Anne, can you please get me a coffee?"

A few minutes later the steaming beverage was sitting on his desk in a white mug with the Zigzag logo imprinted on it. As he sipped his coffee, he started to make a list of the clients he needed to contact in the next few days.

When lunchtime came he dialed the number of the school.

"Can I please speak to Jenny Crichton?"

"One moment, please. Jenny!" It took a moment before he could hear her voice come on line.

"I went to the clinic today." she said in a hushed tone, not wanting the others in the teachers' room to hear her.

"That's right," he said, remembering that she was

# Jenny Can't Get What She Wants

due for a check-up. "And?" he asked, though he wasn't really very curious. Jenny had had some tests done to find out why she wasn't pregnant yet and the results were due in today.

Jenny wanted a family, and felt urgent about having children; Dick couldn't see what the rush was and would be quite happy to wait a few years before taking on the responsibility of supporting children. But if Jenny wanted babies, that was all right with him, too.

"They want to see you," she said in a muffled voice.

"Me? Why?" He suddenly felt indignant, as if his masculinity was being called into question.

"I don't know," she said in that helpless voice of hers. She really wasn't very wise in the ways of the world or in matters pertaining to her own body and systems. "I think they want to run a few tests," she said, almost apologetically.

"Well, I hope you haven't made an appointment for me," he said, already feeling disappointed and frustrated by the whole thing, as he was correctly anticipating the answer.

"I did," it came on cue, fulfilling his worst expectations.

"Jenny," he said with a scolding voice. "When for?"

"Tomorrow. Ten to twelve."

"Oh, no, I can't. I'm busy. We've got a new line out and I have appointments lined up back to back all day tomorrow."

"You have to. It's about our future and if we don't do anything about it now, it might be too late!" She was really sounding pathetic now and when she spoke like that, he knew it was pointless to argue. She would have her way and he knew it.

"All right, I'll go," he said, feeling the grumps coming on. His mind was racing to the need to cancel his lunch with Carl, a friend and a client who owned a chain of fabric stores. He'll tell him it's an emergency. Carl won't mind. After all, he is a friend.

They were sitting in the doctor's office and he was not happy at all. Jenny seemed nervous, too. Doctor McNeil was explaining. "We can't find anything wrong with your wife's cycle, Mr. Crichton," she said, folding her hands in front of her on her desk. "Her egg production seems to be normal."

Jenny looked pleased with herself, as if she personally had contributed to the production of those eggs, which, as Dick knew well, she was born with. This much he knew about the female reproductive system. After all, Jenny had bought a couple of books on the subject and he had briefly studied them when she was out.

"So we need to run some tests on you, Mr. Crichton."

"What do you mean, on me?" he asked, not comprehending.

"Do I need to spell it out? We need a sperm sample." She leaned over and produced a sample bottle from a drawer in her desk. "You can take it home and bring it in tomorrow," she said, as if she was asking him for something as straightforward as a urine sample, "or you can do it here; that is, next door."

He shifted his weight, still unclear as to what was

happening. He felt manipulated and vulnerable; not his favorite position.

"What do you mean, next door?" he asked.

"We have magazines and videos in there," she said, still very matter-of-fact. After all, she was a doctor. She then added, in response to his ill-concealed grimace of distaste, "Or, as I said, you can take it home."

"I think I understand," he finally replied, reaching for the bottle with a resigned look on his face, like a prisoner in the docks after the sentence has been passed. "It's all right, I'll take it home and bring it back tomorrow. Thank you, doctor. Come on, Jenny."

From his tone of voice she knew she would be hearing from him in the car. Sure enough, once safely behind the wheel, he exploded. "What is she implying? That I am incapable?"

"She's not implying anything. She's just doing her job. She needs to cover all possibilities and find out what is wrong. Has it occurred to you that maybe, just maybe you are incapable of having children?' She then quickly added, "Which does not mean that you are incapable of having sex! We all know you are capable of that!" He caught the irony in her tone, but chose to ignore it.

"Well, what do you think?" he asked.

"I don't know. How can I know? But I want to find out the truth. I want to know why I haven't been able to get pregnant. God knows, we've been trying." Her voice softened and there was a gentleness with her now which he felt was contagious as it started to calm him down.

"And when you find out, then what? Shall we be visiting a sperm bank next?"

"I hope it won't come to that, but I think we need to take into consideration all eventualities. There are many things that can be wrong and can be cured. Let's wait for the results before we jump to conclusions or argue about this."

He noticed the contradiction in her statement and was no longer sure whether she wanted to discuss the matter further or wait for the results of the test. He felt suddenly tired of the whole thing and wanted to placate Jenny, so he could be left alone.

"All right, all right. I just get so angry when you take the initiative and put me in a situation where I have no choice." It sounded more like a plea than an accusation.

"Of course you have choice, you always have choice. You can throw the bottle away. You can stop the car, get out and never come back. Okay, well maybe I'm being a bit dramatic," she added when she noticed his "I'm not amused" sideway glance at her. "But you do have a choice. It's your decision. Yes, I brought you to this point and I apologize, if you feel betrayed or manipulated. Now it's up to you."

"I just wish you would talk to me before you start organizing my life for me." He sounded more resigned than rebellious.

"But we are talking—we're talking now, aren't we?"

"I guess so."

"I just want to do what's best for us."

"What's best for you, you mean."

"No, I'm thinking of OUR future and the future of our future family."

Silence fell between them as Dick drove on. A few

moments later he said, much calmer now, "I just don't understand. There are over nine billion people in the world today and you want to give birth to more. The world doesn't need any more children. Not when there is still poverty in the world and not when it is taking all this trouble to conceive. I just don't understand."

"But, darling, we've always said we would have children.

"Yes, some day."

"That's right, some day. Some day has arrived. It's here now."

"Not necessarily. Some day could be tomorrow or in a year's time or in fact maybe even never."

Again they fell silent and he knew, feeling acutely the specimen bottle in his pocket, that he would not be able to get away with it and that, like it or not, the pressure was on to carry the experiment through and find out what was really wrong, if anything.

# CHAPTER TWO
## Dick Hears the Truth

Dr. McNeil looked at them and smiled a professional smile, one that you just knew she had used at least hundreds of times before in similar, or perhaps not so similar circumstances.

"It is you, Mr. Crichton. We have determined beyond a shadow of a doubt that Jenny cannot conceive because your sperm are deficient.

"What do you mean, deficient?" he asked, almost hurt and offended.

"Well, it seems to be a condition that is not described in any textbook. At least not yet."

"Oh, let me guess," he said, thinking as quick as lightning, that Dr. McNeil would probably write it up and be this year's candidate for the Nobel prize in medicine. He was getting positively indignant now and somewhat defensive, too. Jenny shifted uncomfortably in her seat. She was getting embarrassed and hoped that Dick would not become too contrary.

"Well, what is it?" he asked after a moment's silence while all three of them were adjusting to the increased tension in the room. He then added, "Haven't

# The Tale of the Tailless Sperm

they learned to swim yet?" This was said with a touch of sarcasm.

"Well, you're quite right there. They can't," Dr. McNeil replied. "They seem to roll around, rather than swim. They don't seem to have any tails."

"What, no tails at all?" He seemed incredulous now.

"No, just a tiny little, barely visible stump. That's it. Apart from that, they're as round as an egg, but, of course, much smaller."

"Yes, I know, it is the smallest cell in the body."

"I see you've done your homework."

"So, what can you do about it?" he asked, hoping there was nothing more to be said. "Can you give me a pill?" He could live with that—a quick fix-it; no mess, no bother. "A tail producing pill," he added and as he said it, he knew it sounded like a ridiculous proposition. He found his mouth turning involuntarily into an ironic half-smile.

"Well, there's nothing like it on the market at the moment. I would like to do some more research on your condition before I suggest the next step.

"Perhaps they would have heard about it in California; they're always ahead of the rest of the nation in these matters. So let's leave it with me for a week or so and I'll phone you to let you know what I find out."

As they stood up to say goodbye, she added, "And whatever the results of the other tests we will be doing, there are always further things we can try." This was addressed more to Jenny, than Dick.

"Like what?" he asked, a little bit worried, his tone slightly aggressive.

"Like artificial insemination, test tube insemination, the sperm bank; and if all else fails, there is always adoption. There are a lot of parentless children out there, just waiting for a good home. So, you see—there are always solutions. Goodbye, and don't lose hope." She said that with such ease that he thought he could not possibly believe her sincerity.

In the car the dialog continued, as strained as ever when the subject of children used to come up.

"How far do you intend to take this? How important is it to have children? Surely, you do not mean to adopt…"

She tried in vain to pacify him. She seemed to be performing more and more pacifying gestures these days; perhaps she should apply to the United Nations for a job. She smiled at the thought.

"Let's take this one step at a time. So much can still happen. If your sperm cannot swim, because they have no tails, surely they can still be transported to the egg or the egg can be transported to them. They don't need a tail to fertilize an egg; it drops off anyway, the moment it penetrates the egg, doesn't it? Apart from the swimming bit, the tail is really a bit of superfluous equipment."

"Well, that makes me feel better straight away. Thanks a lot!"

"There's no need to become sarcastic. You'll see, we'll find a way."

Dick had major misgivings and did not want to think about it. After all, he was going to play tennis with Carl in an hour's time and that was far more important than arguing with Jenny over the children they did not have and may likely never have.

# The Tale of the Tailless Sperm

So he dropped her off at home and drove to the club, trying hard to think of other things as he passed the familiar streets and turned into the club's parking lot.

Yes, it would be nice to have a son that he could teach tennis to and introduce to his friends at the club. That would be nice. Carl sometimes brought his son with him and Dick had noticed how the boy looked up to his father, always asking him questions and seeking his attention. A boy would carry on his name; he would be like him and he could teach his son how to sell, passing on the techniques that he had accumulated over the years. Dick firmly believed that if you knew how to sell (and it doesn't matter what), you could sell anything to anybody. You would never be poor, jobless or stranded in any way, at least not as far as the material worlds were concerned. A daughter would be all right, too, but no doubt she would take after Jenny and help in the kitchen; she would play with dolls and dress up in pretty starched skirts with colorful ribbons in her hair, just like his sisters used to do. On the other hand, she might learn to play tennis, too. She might ask him to read bedtime stories to her. Yes, a girl would be fine, too.

He felt much better as he locked the car and walked into the club house. He was convinced that sperm did not really need to swim. The solution was clear: if the sperm won't go to the egg, the egg will have to go to the sperm. This was such a simple, elegant solution, it almost made him laugh out loud. Artificial insemination was done all over the world all the time. It was safe and easy; there were no risks involved. So what's the fuss all about? Perhaps his were a new breed of progressive sperm, representing a new generation that no longer needed to make the effort of the long swim toward the egg. Perhaps they knew something and contained new genetic material; perhaps human technology had progressed to the point where the sperm could save their energy to produce a better quality embryo. Yes, that's it. His children will be better, brighter, geniuses perhaps, all because his sperm had economized and evolved to the point that they were able to save all the energy that would have been used up in the production of that superfluous tail.

Is not the loss of a tail the mark of progress? Humans have no tails, even though they used to—they lost those appendages in the course of evolution as they progressed from ape to Homo erectus. Yes, no tail means progress. The human doesn't need a tail to swim, so why should a sperm? The development has been long overdue, marking the next stage of human evolution.

He felt his energy increase, ready for the challenge of a game of tennis, the most sophisticated sport in the world—what he considered to be the pastime of kings, millionaires and, of course, salesmen; successful salesmen, that is.

He met Carl in the cloakroom and they exchanged greetings, as they changed into their tennis whites. As they walked out onto the tennis court, they passed the double doors leading to the swimming pool area.

As Dick mechanically turned his head and looked through the glass in the doors, he saw a children's swimming lesson in progress. There were about twenty of them in the pool and they were having a race. Dick smiled to himself, as he thought that those kids did not

# The Tale of the Tailless Sperm

need a tail to learn how to swim.

"Carl, can I ask you a question?" He turned to his partner.

"Sure, shoot," replied Carl.

"Do you intend to have any more children?"

"Well, we wanted to. Carol certainly wanted a girl. And we've been trying for ten years now, but nothing has happened."

"Really? Well, did you go to see a doctor?"

"No, I figure that if it's supposed to happen, then it will. I'm not going through all their tests and Carol doesn't want to either. Anyway, John is a fine boy and perhaps it is best to quit while you're winning."

"I see."

They had reached the courts and proceeded to occupy their opposite sides. They knew who was serving first—it was Dick's turn to start the game.

# CHAPTER THREE

# Where Have All the Tails Gone?

The man sitting in front of Elizabeth McNeil was no longer young. He seemed quiet spoken and gentle-mannered. He had come alone, and this was already unusual. In her work she had mainly dealt with either couples or women on their own who would come to her looking for information or reassurance, or both. The man wanted a child. He was getting older and didn't know how much longer he had to produce a healthy offspring whom he could guide into adulthood with the kind of vigor and participation he knew was needed to bring up a child. He had come alone because he wanted to know whether there was something wrong with him. He had been married before and the marriage had produced no children. He felt it was down to him, because his previous wife was now happily married with two children, but his own wife of three years had not been able to conceive.

"We need a sample," she said, as she pulled out a jar from her desk drawer with a label on it. The man in front of her—Mr Grzegorzewski—smiled and took the jar from her. "Can you do it now?" He nodded.

# The Tale of the Tailless Sperm

After the man was gone, as she looked through the microscope, she was astonished. Here was another man with the same symptom, so soon after she had diagnosed the syndrome in Dick's case. She wondered whether she was witnessing the beginning of an epidemic.

Jenny and Dick sat in front of doctor McNeil. He had taken his time and had procrastinated enough, delayed enough, postponed enough.

"Okay, let's do it," he said in a resigned sort of voice, but also feeling relief at the same time that the decision had been made and that finally Jenny's nagging would cease. He had agreed to take the next step and here he would draw the line. Artificial insemination—they would take an egg from Jenny and his sperm and mix them together in a test tube, hoping that in that environment even his tailless sperm would penetrate the egg, because it would not need to travel.

Jenny was beaming, as if she were already pregnant. He just did not understand her. There were so many homeless orphans in the world, just waiting to be adopted. But no, she wanted one of her own with her (and his) DNA, with her (and his) genetic traits, with her (and his) intelligence, qualities, characteristics. "Oh well, if this doesn't work, there is always adoption," he thought. The process would take time. Jenny's cycle needed to be monitored and the right time had to be decided and correctly pinpointed. His sperm they already had and the doctor were quite positive it would work. If not straight away, then perhaps by the second or third attempt made.

Doctor McNeil was very positive, very reassuring. That was her job, that was her training.

"I assure you, we shall do everything in our power to make it work," she was saying, looking mostly at Jenny and smiling reassuringly, as she played with a pen that had been lying on her desk.

"Science has taken great strides recently," she added, "and there are new methods of doing this. We have in our laboratory a very advanced culture that boosts the insemination process." Now she was sounding more like a teacher, lecturing to her class. "The success rate has been very high." She got up and came out from behind her desk. "I'll let you know when to come in next." She turned to Jenny. "It shouldn't take long." She held out her hand. "Good luck," she said as they shook hands. Dick felt like he was being treated as if he was a little bit less of a man. Nothing was said, but he wanted to get out of there quickly. Perhaps it was nothing, perhaps he was just imagining that the two women were looking at him with concern. Perhaps it was only in his head.

As the door closed behind Dick and Jenny, Elizabeth McNeil sat down in her chair. She needed a moment to think. She had specialized in fertilization and had written her Ph.D. on artificial insemination. She knew everything there was to know on the subject. This was her life—bringing together the largest cell of the female body with the smallest cell of the male body to procreate new humans. She saw this to be an honorable task and she took great pleasure in the satisfaction of couples who were able to have children thanks to her efforts. But this was something new. She had never seen this condition before. Perhaps she could write it up for the *New Medical Journal*. But to do so she needed more information.

And then she remembered the man who had come in that very morning with an identical complaint. She hadn't told him yet about his tailless sperm, and she was the only one who had seen his sample under the microscope that morning. She went through some files on her desk. "What was his name? Ah, here it is. Gregory. Gregory Grzegorzewski." Completely different history, different genetic, different origin. And yet his sperm looked remarkably like Dick's. Little dots that vibrated and gyrated, like a good sperm should, but only on the spot. No tail, no propulsion, no forward movement at all. The poor things were literally going round in circles, as if chasing their own, non-existent tail.

She couldn't wait any longer. She had to check it out. She turned to her computer and activated the Internet search engine. She knew all the relevant addresses by heart. Anyway, they were all listed as "book marks" on her desktop: medical journals, university research departments, laboratories, clinics and hospitals. She went through them methodically, one by one, sacrificing her lunch break as she did so She knew what she was looking for and her instinct told her it was out there.

Yes, there it was: a French researcher, Dr. Jean Legrande had found the same symptom in one of his patients. He was asking whether there was anyone out there who had encountered anything like it. He had already given it a name: the Legrande Syndrome. She had to smile to herself as she remembered the history of the discovery of the AIDS virus. 'These Frenchmen are megalomaniacs,' she thought, generalizing, as was her very annoying habit. She typed a brief note in response to his plea: yes, there are two cases, right here, in New York. From Dr. Legrande's description the cases seemed exactly the same—no tail, no ability to swim. Just a little wriggle. "I could call it the Wriggling Sperm," she thought, liking better her name to his.

Wondering what could possibly cause this condition, she was asking herself whether it could be a virus or a genetic defect. Wouldn't it be great to discover the virus before the Frenchman found it? But she knew that in her rather primitive laboratory she was out of her depth. Perhaps the Center for Disease Control in Atlanta had the equipment, the powerful electronic microscopes and computers to carry the research forward. Perhaps she should let them know now.

No, it was too early to tell. She would wait and do some further research herself. Above all, she was eager to try the artificial insemination idea and see whether Dick's sperm were totally incapable of fertilization or just reluctant to travel.

## CHAPTER FOUR

# The Tale of Two Non-Existent Tails

It was early in the morning when the phone rang. Elizabeth McNeil rolled over, as she looked at the alarm clock on her bedside table. It was only five o'clock and still dark outside. She sensed an emergency and picked up the phone.

"Yes?" she said in a still very sleepy voice. To her surprise she did not recognize the voice with the French accent that responded to her enquiry. "Bonjour, good morning. Ah, I am sorry to disturb your morning sleep. But here it is way past breakfast time, so I thought it would be a good time to phone."

"Who is this?" she asked, convinced that it was a wrong number.

"It is Doctor Legrande," he said, as if the entire world had heard his name. "And you left me a little message on my web site yesterday, *n'est pas*? And I am coming over to see your specimens. In fact my plane leaves in two hours, so I shall be there around five. Can you pick me up from the airport? I'll give you the flight details."

She couldn't believe it. He was actually coming over all the way from France just to see Dick and

# The Tale of the Tailless Sperm

Gregory's sperm! My God, these Frenchmen acted fast when they could smell a possible ego boost. She sat up in her bed and suddenly she was wide awake. For a moment her ego spoke to her as she briefly imagined her name coupled with that of this Legrande: the Legrande-McNeil syndrome! She shook her head as if trying to get rid of an intrusive fly. She wanted to protest at least a little at first, but found herself saying, "Yes, I'll be there. Give me the details." She had a pen and pad of paper ready by her bed, always prepared to write down her dreams or her thoughts. Her best ideas always seemed to come in the morning, while she was still horizontal.

"I'll be there," she said again reassuringly, as she jotted down the details. "How will I recognize you?" she then asked.

"I will be the handsome Frenchman with greying hair and specially for you, I shall wear a red football cap. I look forward to meeting you, Elizabeth. Goodbye."

She sat for a minute, still holding the receiver, dumfounded. Well, there's a change. Her mind was racing as she tried to remember her schedule for the day. As far as her memory could stretch, she would finish work today at four. That meant she could just make it to the airport in time. There was no point in trying to go back to sleep. She got up at this unusually early hour and made herself some coffee. Pleased that there were still two hours before she needed to get ready for work, she sat down at her kitchen table, attempting to gather her thoughts. She felt something important was happening. A seemingly unimportant and unusual aberration was perhaps turning out to be a global appearance of a new pathology. Well, perhaps not global, but certainly international.

She sat there for a while, thinking about it. She could not remember when there was such an incident before. All previous abnormalities could be found in the books, in magazines, in libraries, or on the Internet, if it was, for example, a newly appearing virus or disease. But this, what was it? Pathology or perhaps development? Regression or evolution on the move? She had no idea and even her instinct was quiet this time. She would simply have to wait and see.

Jenny was scared. She didn't really know what it meant and what exactly the procedure entailed. Doctor McNeil had not been much help in explaining how it worked, probably assuming that she already knew. And she was too reticent to ask. But now she felt she really needed to know and she hesitated to ask Dick, because she felt that he had been too reluctant all along to subject him to her questions and doubts. Besides, if he knew how ignorant she really was, he would surely not allow her to go through with it. So she needed to find out and the best way to do so was to go to the library. She had already read a couple of articles she had Googled on her computer, but she wanted to conduct a more in-depth research before committing herself to a new procedure. So after Jenny finished work work, she went to the reference library, in search of more information about artificial insemination. She remembered having read about it in *Time* magazine and in other journals; she remembered having seen several photographs of healthy babies conceived by this method.

# The Tale of the Tailless Sperm

She knew it had been tested and had been declared safe and reliable. So why was she scared? Not knowing exactly how it worked scared her and she needed to know.

The library was automated and she was glad she did not have to ask a librarian for assistance. All she had to do was to look it up on the computer and retrieve the relevant information. She typed in "artificial insemination" and a whole screen-full of titles of books and articles appeared in front of her. Luckily, there was no shortage of information for her to choose from.

When she located the magazines she was looking for, she realized she had much too much material to go through in one short afternoon. It was difficult to find a place to begin as everything seemed important to her, but she decided to look to the methodology involved and proceeded to read the one article she thought was the most relevant. "The sperm of the donor is inserted by means of a catheter into the woman's reproductive tract," she read and already this was what she needed to know, though she now needed a dictionary as well to look up the word "catheter." She went to the reference section and took the Concise Oxford Dictionary from the shelf. There she found a further description of "a tube for insertion into a body cavity for introducing or removing fluid." She now began to worry that if Dick's sperm were inserted into her reproductive tract, could they find their way to her egg without a tail? Perhaps harvesting her eggs and having the tailless little sperm inseminate her egg in a test tube would be a better option. Now she felt worried about having her eggs removed from her ovaries. Jenny remembered Kourtney Kardashian showing on television her scar after her eggs had been harvested in case she still wanted to have children in the future. The scar was small, but Jenny still worried how painful and how effective the procedure could be.

There was no mistake. The red football cap marked him out from the crowd. As soon as they met, she could feel his urgency, his energy, and his enthusiasm.

"I've got twelve cases on record," he said almost as soon as the greetings were over and they were walking toward the car park and her car. "Seven are French, but the rest are from other European countries—Spain, Italy, Denmark and Sweden. There must be more, but I haven't got to them yet. Yours are the first in North America."

"Wait a minute," she said, interrupting his flow, "you haven't seen them yet."

They were standing on the escalator leading toward the car park and he was facing her, standing on the stair beneath her; her back was facing the direction they were traveling in, so that they could see eye to eye. He smiled; he had seemed to slow down and his accent was more French than before and, she had to admit, more charming than before. "Ah, but I have a good intuition," he said. "I know it is the same thing. I can feel it in my bones. In fact, I can feel it all over." She turned away from him to hide a smile, just as they had arrived at the top. A brief moment went by as they started walking across the car park and then he continued in a more conspiratorial tone, as if he was letting her in on a secret.

"Listen, something is going on. We haven't seen the

end of it yet. This is only the beginning. It's not like SARS where everything exploded onto the world scene and then fizzled out; you'll see. And it isn't like COVID-19 when the pandemic dragged on with second, third and fourth spikes traveling from country to country. I wouldn't have come all this way to look at two isolated cases. This is history in the making. I know it!"

There was no contradicting him, so she let him carry on, curious where he was leading. They were now walking side by side, with him pulling his suitcase on wheels behind him.

"We must be vigilant. We must document all cases. We must create an information network. I have a colleague standing by in China and another one in Ghana, ready to monitor Asia and Africa."

"You really do think this is a worldwide phenomenon, don't you?" she finally asked. They were approaching her car.

"I don't think; I know." She opened the trunk and he lifted his suitcase into it.

"I'll take you to your hotel," she said, but he protested. He would have no rest and seemed as excited as a child at a fairground.

"No, let's go to your clinic first. I must be able to see them."

"All right," she said, already feeling a bit weary of her French visitor.

He seemed glued to the microscope forever, checking her samples, looking at his notes, pulling out his own phials, bottles and jars, of which he had several. She waited patiently, at first standing behind him, in the hope that he might invite her into his deliberations and show her some of his own findings. Then she went to the small kitchenette at the end of the corridor and poured two cups of coffee from the pot which had been prepared earlier by her assistant. She brought both mugs into her laboratory and handed him one of them—the one with the blue rim that she kept for visitors. He gestured for her to put it on the table; he seemed totally absorbed in what he was looking at. At last he lifted his head away from the microscope, picked up the still steaming mug of coffee and stood back from the table.

"This is incredible," he said with an accent on the penultimate syllable.

"What is it?" she asked, not able to contain her curiosity any longer.

"They're all the same," he said, gesturing for her to have a look into the microscope. He added: "No tails. Every one of them. Identical, as if they were brothers and sisters, from the same family." She peered through the glass and there they were—what was quickly becoming a familiar sight. No tails. They looked like little globules with tiny stumps. The sample was still alive and they were moving, but in no particular direction; they seemed to be going around in circles, with their little rump stumps vibrating vigorously. They looked like a caricature of the real thing. She removed the sample and took another one from the surface of the table, where he had laid them out neatly, with their labels showing the country they were from and their coded number. Again she looked into the microscope and again she saw exactly what she had seen before.

"Yes, I wouldn't be able to tell them apart," she confessed, stepping back from the microscope and picking up her mug of coffee from the filing cabinet behind her.

"No, neither would I. Neither would I." He hesitated for a moment, then asked: "Do you think this is the beginning of an epidemic or even a pandemic?"

"I don't know," she replied. "What do you think?"

"I think it is possible," again accentuating the penultimate syllable. "In my experience, if something occurs in two parts of the world, like a new virus or deformity, chances are it will appear elsewhere as well."

"Well, in that case, we should check it out." She was a woman of action and an adept user of the Internet. "Let's find out," she added, as she approached her desk and reached down to switch on her computer. The familiar sound assured her that it was working and as it was getting started, she turned to Legrande and asked, "How should we conduct the search? What are the significant words we are looking for?"

"Ah, that's exactly the problem," he was quick to respond. "It hasn't got a name yet. It needs a name. And it needs to be registered fast, before anyone else claims it. We need to write this up and publish an article in *The Lancet* or some other reputable medical publication." He seemed to be going off on a track all his own.

"Yes, but what shall we look for now?" she asked again, with increasing urgency in her voice.

"I don't know…" his voice was slowing down as he pondered the next issue which appeared in front of his

# The Tale of Two Non-Existent Tails

mind. He walked up to the window and looked out at the courtyard below.

"We should call it the Legrande Syndrome. After all, I spotted it first."

"You already gave it a name," she reminded him, as she got up from the computer and put her hands in her white laboratory coat pockets. "I found it on the Internet, remember?" She was getting slightly annoyed at this unexpected delay.

"Ah, this is true," he said in a slightly dreamy voice. No doubt he was picturing himself receiving the Nobel prize for medicine, she hypothesized. "Come back from Stockholm, doctor," she said, expecting him to understand her train of thought.

"What?" he turned away from the window, facing her at last. He had not understood.

"Are you asking about the other cases?" he asked, and then added, "No, they were not in Stockholm. I had one case in Malmö, two in Copenhagen, one in Malaga, Spain and one in a little village in Tuscany. That's it. Together with the seven French samples from Paris, Toulouse, Reims and Boulogne, that makes twelve. Just twelve and quite dispersed, with no easily determinable connection. I have visited all those places and have taken fresh samples from each of the men concerned. They are all here, in my little portable, battery operated freeze box. Specially designed for the purpose." He pointed to his bag which was sitting on a chair next to the laboratory table.

"Okay, so let's assume that this is something that is spreading. Perhaps the name The Legrande Syndrome has already been adopted by other physicians in other countries. Let's check it out." She approached her desk again, where the computer was now humming and ready to spring into action. She started typing and soon there appeared upon the screen the message that Jean Legrande had sent out some days ago, requesting any information about sperm without tails, manifestations of The Legrande Syndrome. There was nothing else under that name. She then tried to match other combinations of words, like "sperm without tails," "tailless sperm" and "conception problems." She was getting further and further away from her target, as she started introducing new words, like "fewer births," "no pregnancy" and "impregnation."

At last she turned to Jean, "This isn't working," she said. He had sat down at the desk and was repacking the phials and jars, occasionally taking a sip of coffee.

"Don't you understand?" he finally said. "It's entirely new. No one has heard of it before. I need to go to a hotel. I must write an article about this before it becomes too widespread. This, my dear doctor, might change the face of the world."

"What do you mean?" she asked, turning away from the computer desk, suddenly alarmed.

"Well, think about it. If this is the beginning of an epidemic, just think where it could lead."

She slumped in her chair, as the logical extension of this situation began to unfold before her mind's eye. A series of pictures began to travel through her brain, one more extreme than the other. First, the population of the world would diminish and everyone would be happy about that, because already there were too many mouths to feed on planet Earth. People in third world countries would be better off as a result, but people in the West, who still thought that having a baby was their inherent right, rather than privilege, would become more and more desperate. More and more funding would become available to overcome the crisis. This was good because as it was, she could never get enough money to pay her staff or to be able to afford decent equipment. Even her computer was by now around three years old and all her applications were out of date already. So people would look for new technologies and she would be inundated with clients. She would need to find a way to help them. Perhaps importing the tailless sperm directly into the egg *in vitro* would work? Well, that was an interesting concept.

Jean had packed his things and was standing next to her desk.

"I am beginning to feel the jet lag," he said. "Let's go to a café and talk this thing over."

"Okay," she said. She had cancelled her two appointments for that afternoon, knowing that Jean would be in town. "Let's go."

There was a Starbucks around the corner from the clinic and they were soon sitting opposite each other with their cappuccinos in front of them.

"This is a disaster," Jean was saying. "Think what this could cause." He was talking about a disaster, but his tone of voice sounded more excited than worried.

"That's precisely what I've been doing for the last twenty minutes," she admitted.

"But you've got to think globally, yes? Just think—

fewer people. It will begin with the babies. A bit like an anti-baby boom. A baby decline. We need to find a name for it. Baby boom was so catchy. How about baby bust? No, that has a double meaning, doesn't it?" He seemed to be obsessed with names and he seemed to be quite satisfied with the little dialog he was having with himself, she observed. "In France already the population is declining and there are fewer babies being born. France is dying! With this the end will be so much closer than we had ever imagined."

It began to sink in. "Babies," she said, thinking about the consequences. "No more nurseries. No more Baby Gap. What will all the nursery teachers and nannies do? Toys R Us will go bankrupt." She thought of her friend Freda who taught two-year-olds to read, using a new computer program based on color, shape and sound.

"I never liked that name anyway," he interrupted in an off-hand kind of way. "That's just the beginning," he continued, undaunted. "After the babies come the toddlers, then the kindergartens and then the schools. They will slowly close down, too."

"Oh, my God!" she said, becoming more and more worried as the conversation progressed. "No more teachers, principals or janitors. All those empty buildings. No more school shoot-outs. Who will the drug dealers sell drugs to?" He looked at her, horrified, as if he wasn't sure whether she was joking or not.

"Never mind them," he said, "they will always manage to find customers. But a whole industry will die. No more school books or papers to mark, and the telephone companies will lose most of their teenage customers. What about the music industry? What about fashion, parties, tattoos, jewelry, make-up? What about car insurance? The whole world will start to change, dramatically. Think of all the empty bedrooms. No need for new houses, except to accommodate immigrants."

They were beginning to understand one another. She continued the thought process. "Cheap lodgings. University dormitories turned into hotels. Colleges become conference centers. Hotels lose customers."

"What will happen to all the school buses you have in this country?" Jean picked up where she left off.

"And finally it will hit the grown ups. The work force will dwindle as the population ages. It will only be a matter of keeping the economy going so that those who are still around can survive."

"There will come a time when more people will be retired than will be working. No doubt pensions will become obsolete and old people will have to continue to work, just to keep the economy going."

"If they are still able to work," Elizabeth added.

"Property prices will plummet. We will each have our pick. We'll be able to live wherever we want."

"As long as we live. But think of a world without children, without laughter. Who will we condescend to? Whom will we teach? Who will think they know better and not listen to us?"

"It sounds like a disaster."

"I told you so. And it will be if we don't do something about it now."

"Before it is too late." They fell silent for a moment.

Each of them was exploring a world without children in their minds.

"And think," Elizabeth finally broke the silence, "what it would be like for that final human being on Earth. How lonely they would be, wandering from place to place, with everywhere to go, but no one to talk to."

"It doesn't make sense. It only makes sense if there is simultaneously a more advanced super human that will take over the world."

She looked at him incredulously. "Do you really think that is possible?" she asked. "Are we missing something? Do you think that alongside the defective sperm without a tail, there might be others emerging, perhaps with two tails? A sort of super sperm that will begin to take over the ova of the world?"

"I don't know. I didn't say that. But it is possible; everything is possible. Maybe not even a super sperm, maybe a super ovum. Maybe enhanced genetic material in both. We have to wait and see. But let's be vigilant because I believe that nature has strong survival instincts.

"Yes, that makes sense. After all, the human has survived this far, against all odds. There have been those prophets of doom that have predicted our demise long ago."

"Yes, we have survived despite wars, and famine, and plagues, and diseases."

"And despite the destruction we have inflicted upon our environment."

"Doctor McNeil, I believe we are the soldiers of tomorrow. It is up to us to ensure the survival of the human race and the promulgation of the species."

"Call me Elizabeth," she said. "Your English is very good, doctor. And I like that French accent. Yes, I do agree."

"Call me Jean," he replied as he thought that at last she was beginning to appreciate his intellectual power. Of course he had known that she would in the end. They all did.

# CHAPTER FIVE
# The Race Toward Success

He worked all night. He set up his portable microscope and his laptop computer and was further comparing samples, as he wrote down his findings. He had taken some of her samples and surreptitiously placed them in a freezer bag for further investigation while she was out of the room making coffee. She would never notice, of course, because, after all, she had millions of them; millions of ineffective little sperm.

He smiled to himself as he started writing his article. It was early days and 14 samples—12 in Europe and two in the United States—could hardly be called an epidemic or a pandemic. Well, not yet, at any rate, though the fact that the cases were so widespread pointed to a curious coincidence, which might turn out to be not a coincidence at all. He felt he had to act fast before anyone else published before him and gave it another name. Someone else might call it something like AIDS—an acronym, for goodness sake! What a missed opportunity to put a name to a discovery. He smiled as he thought of possible acronyms and initialisms, like TSS for Tailless Sperm Syndrome or JLS for the Jean Legrande Syndrome, or even JLEMcNS for the Jean Legande/Elizabeth McNeil Syndrome, though this last initialism seemed unnecessarily generous. He decided there was no need to include Elizabeth in his thinking.

He had enjoyed their conversation; he felt that they understood each other and that she would not object to his article. Sure, he would mention her and give credit where credit was due, but after all, he had collected twelve samples, traveling around Europe at his own expense, whereas she had produced only two, both of which were from men who had walked into her clinic unsolicited. There was no comparison, no competition, and he was sure that she would see that and appreciate his point of view. After all, she seemed like quite a reasonable person.

And so the article was written. He explained the condition in detail, describing the eight locations and mentioning the doctors involved in each case. He would get someone to take a photograph and then his work would be complete. He then checked his e-mail, satisfied with the results so far. There were several messages for him, including one from Florida and one from Tennessee. As he did not recognize the sender of either of these messages, he opened them first and, sure enough, they were from doctors—one an obstetrician and one a gynecologist, sending in the results of their findings. They, too, had seen the same syndrome appear, when checking out the spermatozoa of their patients' partners.

"So, I must go to Tennessee and to Florida," he thought, as he pulled out his flight itinerary and checked his reservation. It was past midnight and he was very tired; he decided he would call the airline in the morning.

Joe Kowalski was a self-made man. He had dropped out of Harvard medical school and never became a doctor. Now he was quite grateful for this turn of fate because he enjoyed his work and thought he was so much better at editing a magazine than dealing with patients. After all, he never did like the smell of hospital wards and he would not look good in a white coat. But the written word, that was something else. He loved language and the fact that it could be shaped and honed into an innumerable variety of ways. He had worked his way up from junior clerk to editor-in-chief and he was proud that he had done so on his own merit, by hard work, persistence, and a stubborn attitude that never took no for an answer. It had served him well, this resolution he had made to make it in a competitive world.

The *New Medical Research Magazine* had started up as a small local information sheet, contributed to by doctors in the field, laboratory technicians and scientists, and it was funded by the advertising incorporated in the magazine as well as by a small grant from a large medical supplies company that from time to time would print articles extolling the virtues of their new products. When Joe started work it was a monthly and the editorial board originally consisted of half a dozen people, as well as a couple of volunteers who would come in to help with the final production touches. Two of the permanent staff were mainly on the phone selling advertising space; one looked after the accounts, one did the lay-out and Fred was the editor-in-chief. Joe came in as office junior, which meant making coffee, going to the bank, being a receptionist and telephone operator, and occasionally being sent out on an editorial assignment, or maybe an interview with a researcher, or a brief report from the hospital, or a description of a new drug. Joe enjoyed this running around and soon had made himself indispensable by being everywhere all at once and poking his nose into everyone else's business, but in a friendly, not malicious way. "That kid has too much energy," Bert, one of the sales representatives, used to say, as he watched him running around the office, always with a smile and a chat for everyone.

"We need a female in this environment," he would add, with a complaint in his voice. He was an older man, in his late fifties, a real veteran of the tele-sales marketing scene. Very dedicated to his work and proud of his results, well organized, almost pedantic. He kept to his schedules and never missed an appointment. Slow, methodical, effective—that was old Bert's way. Well, he got his wish in the end, but it was too late for him to enjoy it, because the woman that came to join the team was in fact his replacement, when it was time for him to retire.

So Joe had made up his mind that this was the place for him. And although he was only twenty-two when he dropped out of medical school, he felt that here he could make a career for himself. From the beginning he had his eyes on the main job—that of editor-in-chief—together with the Mercedes Benz, trips abroad and well-cut suits that seemed to go with the job. The money and the prestige also seemed to go with the job. To get there he knew he had a lot to learn, but he did not mind how long it would take; he was determined to find out exactly how the office was organized and what made it hum. For it did hum—there was a good chemistry among the men there, with banter, camaraderie and the willingness to fill in for each other when one of them fell sick or needed some time off. Sure, there was competition amongst them, but by Friday lunchtime they were ready for their trip down to the local bar for a beer together, leaving one of them behind to answer the phone and deal with any emergency that might crop up. Perhaps once or twice a year there would be a company camping outing somewhere in the wilderness of up-state New York—walking several miles a day, canoeing and swimming. And then, of course, there was the compulsory Christmas party, when they would bring their friends, wives and girlfriends and get drunk together, dancing, eating and exchanging numerous toasts and occasional gifts. The boss was not a miser and for some of them this was their only opportunity to frequent an expensive restaurant and to wear an evening suit.

They had been getting busier and busier, with the news sheet expanding to become a prestigious magazine. Then all this male bonding changed when Maggie joined the team. Maggie was employed as a receptionist and general assistant, so she took on some of Joe's duties and thereby released him to become more of a journalist and less of an office junior, which suited him just fine.

The day Maggie walked into the office, everything changed. She was young, beautiful, intelligent, vigorous. All heads turned at once in this previously male environment, with pens held mid-air, telephone receivers being automatically returned to their cradles, and Joe holding the coffee pot, getting ready to pour the first round of coffee of the day.

"I am Maggie," she announced, to all and no one in particular, as she stood there, in front of the door, looking around the office. The boss was the first to react. He got up from behind his desk and walked into the middle of the office, where everyone could see him. He then proceeded to speak to all of them, who were still frozen in their respective postures.

"This is Maggie," he said, although they obviously already knew that. "Maggie is joining the team as of today." Those words seemed to release the spell and Joe put the coffee pot down, and the others shifted in their seats, much like a video that had been released from its pause button. "She will be our new receptionist and will take all incoming calls. She will also deal with correspondence and banking." He took a few steps toward Maggie, who was still standing close to the door. "Welcome Maggie," he said and held out his hand. After they shook hands he started to guide her toward the receptionist's desk where the telephone system was located as well. "So why don't you sit down and make yourself at home, and Joe will show you the ropes. Joe, can you please come into my office?" His office was not really an office, but a couple of screens, a desk and a filing cabinet, as the entire office was open plan, including the conference "room," where each new issue of the magazine was planned behind a series of screens attached to two corner walls of the building. Joe followed the boss and sat opposite him in front of his desk.

"This magazine of ours is expanding, Joe," the boss said with a smile as he looked directly at Joe. "You've been here how long, two years?"

"Yes, it will be three years in February," Joe replied.

"Well then, it's time for you to move on, too. I want you to fill Maggie in on all your office duties—show her the office, introduce her to the guys and show her the telephone system. Then take her through the day to day procedures, like banking and opening the mail, as well as answering the phone and using the computer. If you sit with her for two days, do you think she can learn all she needs to know to fulfill the duties of a receptionist and junior clerk?"

"Sure, I'm sure she can manage that."

"Well then, I want you to drop all those clerical functions, except when Maggie will be ill or away on holiday, and take on the job of junior editor. Do you think you can manage that? (The accent was on the word *that*.)

Joe sat up straight in the chair. "Yes, of course!" he exclaimed. That's what I've always wanted to do!"

"All right then, consider it done." Silence fell for a moment as Joe hesitated. It was clear that he wanted to say something, or ask something, because he was now looking at his shoes, but not moving from his chair. The boss asked, "What is it, Joe? Did I forget anything?"

"Yes, I think so." Joe lifted his head. His cheeks were red and his voice was muffled when he asked, "Does this mean that I will be getting an increase in salary?" he finally blurted out of himself, struggling for the courage. He knew that if he didn't ask now, the boss would say nothing and his hand-to-mouth situation would continue. The boss was quick to answer, obviously having already thought it through.

"Let's wait and see. See how it works out. Let's give it two months during this transition time. There's a lot for you to learn. I'll assign you to Bob and he can teach you how to go about your new duties." He got up from behind his desk, came over to Joe and patted him on the shoulders.

"Come on, don't be disappointed. It will all work out. Come with me and I'll tell everyone about your new position."

Joe felt disappointed and elated, both at the same time; he couldn't work out which feeling was stronger in him. He felt his career was progressing and his ambitions were slowly being fulfilled. He was very glad to be released from his duties as receptionist, junior clerk and office manager. He got up and followed the boss back into the general office. As they emerged from behind the screens, he caught a glimpse of Maggie's legs beneath the receptionist's desk. She was speaking on the phone and she had crossed her legs with one shoe dangling off her foot, held in the air by its flimsy connection to her toes.

"One moment, please." She looked up from her desk, where her line of vision had rested, while she was listening to someone speaking, holding the receiver with one hand and playing with a loose strand of auburn hair with the other. She now covered the mouthpiece with her other hand and looked at the boss questioningly.

"Who is Bob and how do I connect him?" she asked with obvious confusion in her voice. Joe quickly sprinted forward and was by her side in a couple of seconds.

"See these switches," he pointed to the dozen switches on the telephone exchange board. "Each represents someone's extension and they are all labeled. Bob is number four—Harris. So you just flick the switch to ring his phone and when he answers, tell him who is on the phone and transfer the caller by flicking the switch in the other direction, that is in the same direction as the switch representing the line that the caller is on. Like so." He demonstrated the action he had described and spoke briefly to Bob, who was already alerted because, having heard his name mentioned by Maggie when she first called out for help, he had been subsequently watching with amusement Joe's little telephone demonstration.

Bob responded, "Okay, got it" and Joe connected him up with the incoming call.

"See, it's not that difficult," Joe said, as he smiled at Maggie. She smiled back and reached for the handset, which he was still holding in his right hand. As she took the handset from him, his fingers brushed against hers and as he released his grip, he felt an unfamiliar fire running through his body, from fingertips to his groin and back up to his brain. As his hand became free, he quickly went over to the gents washroom, to cool off and splash cold water over his face.

He was gone for a couple of minutes, and when he returned, the boss was still standing in the middle of the office, and had already started making his announcement, for all to hear. "So Joe is going to become junior editor from today and I want you all to support him and help him with his new duties. I would like you, Bob, to take him under your wings for a month or two, so he can learn how to become a writer and a real journalist." The five men, who were listening to this little speech, spontaneously burst into applause. Maggie joined in as well. Joe took a small bow. Bob was standing in front of the screen that separated his desk from the receptionist's area and, having completed his call, he was now surveying the scene with curiosity. As the applause subsided, he responded, "That's fine with me, boss, as long as he doesn't slow me down."

"He won't," the boss said, "and if he does, let me know. We don't have any time to lose. The next issue planning meeting is next Monday and you had better all be ready with your proposals and suggestions."

"Yes, boss"—a chorus of mutterings went around the office and all went back to what they had been doing before the interruption.

Legrande knew all about the history of the *New Medical Research Magazine—NMR* for short—and how it became known in the world, gaining recognition and prestige. To publish in *NMR* was synonymous with becoming known in medical research circles and he knew it. All major articles were posted on their website and any serious researcher or Ph.D. student always looked there first. They had a first rate bibliography and there was never any need to obtain permission to reprint—for a small fee permission was automatically granted, no questions asked. Legrande had met Joe before at one of the medical conferences he attended in Miami, so he thought that Joe would be the best person to contact, once he had finished his article. He needed to register his name with this new syndrome, because all his instincts told him that this could be very important, not only for his career, but possibly for the future of the human race.

He dialed the number and asked to speak with Joe.

"Hello," he said, not pronouncing the *h*. "This is Jean Legrande."

"Ah, Jean, how are you? How is it going?"

"Joe, I have a bomb." After a moment of silence he realized he wasn't getting through. "I mean something important, something big. Can you meet me today? I need to travel to Miami tonight."

"Today? That's short notice." Joe looked at his diary. "I can fit you in at four, but it better be good."

"I don't want to come to the office. Too many ears can hear there. Let's meet for drinks at that bar across the street."

"All right. I'll be there." Joe hung up and thought, "Those Frenchmen, always exaggerating and always ready for a drink." He did not like to drink during the day and his work day seldom finished before eight in the evening.

The bar was quite busy, even though it was only four o'clock. He ordered a tomato juice and went to sit down at a corner table where it was comparatively quiet, when Jean walked in. He had his briefcase in his right hand, while he proceeded to take his sunglasses off with his left. Joe recognized Jean at once. He noticed that the Frenchman looked urgent and older than Joe had remembered him, but still fit and vigorous. Joe walked up to him with a smile.

"Ah, you're drinking a Bloody Mary," Jean said, indicating the glass Joe had left on the table.

"No, a Virgin Mary," Joe replied and then he added, as he noted Jean's lack of comprehension. "No alcohol, just Worcester sauce."

"Ah, I see." They sat down together and Jean placed his briefcase on the floor next to his chair.

"What can I get you?" Joe asked.

"A glass of white wine, please." Joe went up to the bar to fulfill the order, while Jean took the pages of his article from his briefcase. He had e-mailed them to his secretary in Paris, who in turn formatted and printed them and faxed them back to him at the hotel. He then copied the fax and stapled a copy, ready for presentation to Joe. As Joe returned to the table, carrying a glass of white wine for Jean, his eyes fell upon the article.

"What is this?" he asked, preferring to hear the explanation directly from Legrande, rather than reading the whole thing.

"It's an article about The Legrande Syndrome," Jean said.

"I can see that," Joe replied, beginning to feel slightly impatient; having put aside his precious time and having spent money on drinks, he at least expected Jean to be cooperative and forthcoming. "But what is it? What is it about?" he added, making allowances for the fact that his interlocutor was French.

"It is something that is just appearing and no one has yet written about it. I checked that out."

# The Race Toward Success

"He wants me to think it's urgent," thought Joe. "All right, let's listen." He sat back in his chair and took a sip of tomato juice.

"It is beginning to manifest itself in the sperm of men around the world—so far in the industrial west. I have personally observed it in Europe—in Sweden, France, Denmark, Italy and Spain. The other cases have been here in the States, but who knows how many there really are around the world; we haven't heard about them yet.

"The sperm are mutating and no longer have any tails, so they can't travel to the egg. This could change the course of history as we know it."

Joe was digesting the information he was receiving. "Now, wait a minute. You've seen this for yourself?"

"Yes, I have. I even have samples and photographs. See for yourself." He opened the article to the relevant page and placed it in front of Joe.

"And these men, are they impotent?" Joe asked.

"That's right. Not one of them has any children. At least not from his own genes."

"What about *in vitro*? Does it work?"

"Well, we haven't tried it yet. It's too early to say. That is a possibility, I guess, but we simply don't know. I just want to alert the medical world to the existence of this syndrome and to the inherent dangers."

"And no doubt make a name for yourself at the same time," thought Joe, but he didn't say anything.

"If it is an epidemic and it becomes widespread, then its effects will no doubt spread faster than the technology to halt its progress. Can you see women in the third world being artificially inseminated *en masse*?"

"Well, perhaps sperm banks will become the most secure money making investment of the future." This was said with a certain amount of sarcasm.

"I'm serious," said Jean. "This could be very far reaching."

There was a moment of silence, and then Jean continued, "Anyway, we won't know for a while, of course. It might take years to progress. But we should publish now. What do you think? Do you want to print the article?"

Joe hesitated, as he picked up the stack of typewritten pages. As he did not reply immediately, Legrande added with a smile, as he sipped his wine, "I can always publish elsewhere," he said.

Joe looked at him with disbelief and replied, "Let me read this and present it to the editorial board. I'll give you my answer by Wednesday."

Elizabeth decided she would try to attempt to fertilize an egg with Dick's sperm. If she could do it in the dish, first by contact, and if that didn't work, then by injecting the sperm directly into the egg, then perhaps she could help Jenny get pregnant, after all.

She was glad Jean had gone off to Miami. She found him to be arrogant, self-centered, big headed and obnoxious, all at the same time. Yes, it was true that they were interested in the same things and that they had had a really good conversation, guessing what the implications of this new discovery might be and envisioning the future without children. But he had been so preoccupied with his own success and career that she felt he didn't really care much about anybody else but himself.

She did care. She wanted to help women have children. That had always been her mission in life. Perhaps because she could not have any herself or simply because she loved children so much—their laughter, their innocence, their potential and the very fact that their future lay ahead of them, still unused, not defined—a wide open opportunity.

She discovered that she was unable to conceive as early as when she was a student at medical school, studying for her exams. She found that her periods had stopped quite abruptly, unexpectedly, without any warning. Her parents sent her to the family doctor who in turn sent her to a specialist. The diagnosis was devastating to her—she had a rare condition, which caused her fallopian tubes to narrow to a point where the egg was not able to flow through to her womb. She was not married then, but the experience helped her decide which branch of medicine to specialize in, and she became a gynecologist. After years of resignation and heartache she now realized that if she found a partner who could be a godfather to her children, she could harvest her eggs and have one impregnated *in vitro*.

She thought again about Jean—if he did have some new cases in Tennessee and Miami, then maybe this condition was becoming widespread, after all. "Those dealing with it are going to need all the information they can get," she thought. So although it was late, she went back to the laboratory and prepared a dish. She had a few eggs that she kept in the fridge, the donor unknown.

She then checked the freezer for Dick's sperm and found the phials filled with samples. "There you are, you little devils," she said to herself as she extracted a small quantity and added it to the dish.

Joe remembered his first assignment as junior editor. He had been so pleased to be sent out into the field with his new mentor, Bob Harris, he really felt he was growing up. They went to visit the first AIDS ward in the local hospital and he was going to help Bob take notes and type up his report. On the way there Bob kept talking about his ideas about AIDS and Joe wasn't sure whether he was being serious or not. Bob firmly believed that the AIDS virus had been manufactured for the purpose of biological warfare in government funded laboratories. He said he had proof, but wouldn't be writing about it in his report, as he needed to go along with the policy of the *New Medical Research Magazine.* Joe was surprised Bob was telling him all this, and at first he didn't quite believe what he was hearing, but then he realized that Bob made no secret of his convictions and ideas.

"Why do you think so many Haitians get AIDS?" he asked.

Joe had no idea. "Because they have been experimenting with genetically targeted diseases; they are learning to genetically engineer viruses so that they will attach themselves to the cells of specific groups of people, influencing their genes and destroying their immune systems. They have the technology."

All this was new to Joe. He felt like he was being educated and brought into the world of intrigue and politics, and somehow he was losing his innocence. He did not know what to think.

Bob continued to believe in many conspiracy theories. When the COVID-19 pandemic swept around the world, Bob was known to believe that it was manufactured by the Chinese government in a laboratory in Wuhan for the purpose of biological welfare.

When they arrived on the new AIDS men's ward of the hospital, another shock awaited him. Young people looking emaciated, ill, wasted, with large, glassy eyes, staring at them as if they were looking from another world. They looked older than their years and some were frightfully thin. They interviewed a few who were still capable of talking and not yet hooked up to machines.

"I know I am dying," a young man said. "There is no cure for me. These pills they give me aren't doing anything for me. I hope there will be a cure soon, so others don't need to suffer as badly as I have done. But it's too late for me."

Joe saw that the young man had been reading a book, which lay open on his bed covers.

"What are you reading?" he asked.

"It's *Cancer Ward* by Solzhenitsyn," he replied. "Bloody real," he added.

That encounter brought Joe face to face with the fact of death and dying. He was looking at a man who was only a few years older than himself and who was facing the fact of his own mortality and the fear of wasting away and degenerating into blindness, or suffering severe brain damage.

"I don't mind dying," he said. "It will be a relief. But I want to go consciously, so I know what is happening to me, so that I can fully experience and feel the process of dying. I want to be surrounded by my family, but they are too scared to visit me. Only my younger sister comes to see me occasionally."

Joe took down some notes and promised to return, to check in on the young man. He asked the man if he wanted anything and in reply the man asked for some notepaper. He explained that neither his parents nor his older brother had visited him since he had become ill and was diagnosed with AIDS. He was feeling lonely and wanted to write to those who had been close to him during his lifetime. Joe promised for the second time that he would return.

Joe was excited. An article was appearing that had his name on it, for the very first time. It's true that the by-line said, "Additional material by Joseph Kowalski," but nevertheless—there it was in print for all to see. He had written up the story of his encounter with Mario on the AIDS ward and Bob was moved by what he had written, and agreed to add it in. He walked into the office the next morning full of expectations and promise, as if everyone would already know about his success and be as happy about it as he was. At the door, Maggie called out his name, "Joe, have you got a moment?" She was looking worried. He walked over to her desk with a smile on his face.

"Yes, Maggie, what is it?" He noticed that his tone was slightly patronizing. He quickly reminded himself that just a few weeks ago he was just a junior office boy.

"Sorry, but I can't remember how to transmit the file to the printer." Their printer was at least a hundred miles away, but the system worked via e-mail and it was a monthly ritual.

"Have you got the pdf file?" he asked.

"Yes, it's all complete and finished." He was very tempted to open the file and have a look at the article layout, but he curbed his curiosity as he turned his attention to the difficulty at hand. The article was called *New AIDS Ward at Local Hospital* and it was the lead article, with photographs and interviews with doctors, nurses, patients and a virus researcher. Joe knew that time was pressing, as he asked Maggie, "Are you on line?"

"Yes, I am. But I can't remember the password and I'm not sure I've got the right address."

"Okay, let me show you." He sat down in her chair; as she stood next to him, watching intently what he was doing, he caught a whiff of the scent of her perfume. He became very aware of her standing next to him. A few key strokes and the job was done, though the computer was taking time to complete the transmission.

"The password is Hermes," he said. "And this is how you do it. I wrote it all down for you," he added, as he pulled out a navy blue binder from the bottom drawer of the filing cabinet located next to her desk.

"It's also on the computer under *Instructions*. But don't worry, all you have to do is ask." He got up to make room for her and found that he was reluctant to leave.

"I'll leave the binder on your desk, so you can have a look at the instructions when you have a spare moment."

# The Tale of the Tailless Sperm

"Thanks, Joe." She looked up at him and smiled sweetly. She won't be here long, he thought to himself, imagining that she would soon get married and leave the office for a different kind of life. Little did he know that he would become instrumental in her leaving.

The editorial board had grown over years. The magazine had gained national as well as international recognition and Joe had grown with it. Looking at them now, as they all filed in to the boardroom, he felt a swelling pride that always seemed to permeate his entire being during their editorial meetings. They had two such meetings every month—one was on or around the 20th, when the final lay-out of the issue was being prepared and agreed two months in advance, and one was around the fifth, when they reviewed the final version of next month's issue for possible improvements and mistakes, and last minute additions. It was also when they began planning the next issue. The system worked well and they managed to keep ahead of the competition and bring to the world current medical news, as well as stories of general human interest. They were always in pursuit of new diseases, new discoveries, new drugs and new cures. Several years ago (Joe couldn't remember exactly, but he thought it was four or five) they had introduced a new controversial section entitled *Alternative Medicine*, which had grown over the months to become a substantial ingredient in their monthly deliberations, giving rise to its own sub-editor-in-chief and small staff. It had changed the nature of the magazine and the office itself, and Joe smiled to himself, as he thought back to the days when their entire staff could fit around the desk of the editor. The new offices were now located in a large building on a busy street, advertising its location to passers by in large neon lights. The advertising section of the magazine was bringing in a large amount of revenue every month and although they still depended on their government grant for continuance, if necessary, they could scale down and survive without it. When the *Alternative Medicine* section first appeared, they were threatened with the loss of their grant, but they fought and won their appeal. The government could simply no longer ignore the growing popularity and credibility of ancient and new branches of medicine. So rather than ignore their existence, making them illegal, discrediting them and closing them down, as they first attempted to do, they had decided to take the other route and not only recognize their existence, but to embrace them and regularize them by issuing a whole new set of rules and regulations that ensured that a large percentage of the magazine's substantial profits would be channeled into government coffers.

New Age and alternative medicine had become a multi-million dollar industry and in fact was fast outgrowing in popularity and dollar value mainstream medicine. Joe wondered whether in fact the term *mainstream* would one day not need to apply to more traditional and experimental methods of healing. Energy work was no longer a weird expression. Reiki, Therapeutic Touch, and Body Talk were fast becoming popular healing modalities. Other systems based on ways of increasing the flow of energy throughout the human body, like various forms of yoga and Tai Chi classes, had now entered many gym schedules and even school curricula.

So there they were—five editors, two sub-editors-in-chief, plus a whole battalion of regular contributors, correspondents and free-lance writers, as well as lay-out artists, photographers, proof readers and secretaries. As he watched them file into the room and take their seats, he was amazed at how life had worked out for him. He could barely get over the fact that here he was, Joe Kowalski, editor-in-chief of a major medical magazine, read by scientists and the general public alike. Somehow he had managed over the years to corner the market and to combine interesting facts and discoveries with scientific research. Perhaps the fact that he had incorporated alternative medicine and real human stories, perhaps the beautiful photographs and innovative lay-out had something to do with it; whatever the reason, he was a success and so were his employees and contributors—most of them household names with books, articles, videos, and films to their credit. Medicine had become such a major field of interest over the last years, no longer the domain of the few who were either specialists or were ill. Everyone's health was in danger with the over-population, the pollution and the rarity and expense associated with organic foods that had not been genetically engineered. Everyone seemed to be interested in preserving and improving their health while new genetically modified diseases were appearing at an alarming rate.

So there he was, ready to review proposals for the next issue, and his was the final word. As they gathered, he started as he always did, by asking, "Well, what have you got for me this month?" This month was October, but the magazine being worked on was for December, with special features for the holidays, as the majority of materials were brought together two months in advance. This then allowed for a last minute insertion into the next month's edition, if that was considered necessary and beneficial.

As they went around the table, Joe listened, adding comments here and suggestions there, but really his secret was that he was not a researcher or a scientist and although he had spent a couple of years in medical school, he had not kept his knowledge current and he did not know a whole lot about modern medicine. He did, however, have an impeccable journalistic instinct and knew how to read the trends of his readers, and the interests of the general public. For specialist articles, he had his associates and researchers, and somehow between them it worked, with the *New Medicine Research Magazine* quickly becoming the most read medical journal in the world.

"Here is something that might interest you." He laid Legrande's article on the table. "The Legrande Syndrome. Has anyone ever heard of it?" He looked around the table and all he could see was inquisitive looks. No one had heard anything.

"Kathy and Ali, please read it and tell me what you think by tomorrow." He turned to his researchers. "We can add it to next month's issue, otherwise someone else might publish first. I've got photographs for it coming as well." Kathy picked up the manuscript. "I'll get you a copy," she turned to Ali and stood up to go to the copy

# The Tale of the Tailless Sperm

machine.

"All right, that's all for today. Well done, everyone. I think we've got ourselves a great December issue."

It was a curious life, in a way always living ahead of himself, already thinking about Christmas and the New Year in October.

As soon as he got into the office and settled in and turned on the computer, beginning to check his e-mails, the phone rang. It was Ali, a regular contributor to the magazine and senior consultant.

"I read the article," he said. Ali was phoning from the hospital where he worked three days a week as surgeon. Educated in India, Ali had done very well in establishing a consulting practice in two major hospitals and operating three days a week. "It sounds interesting. I think it's a bit risky in that it seems to me to blow a small discovery out of proportion. But then if this Legrande fellow is right, you will have wished you published first."

"So the bottom line is, is it worth the risk?" Joe trusted Ali's judgement; he had relied on it before, with positive results.

"Yes, I think it is," Ali replied. "The man has done his homework and he is obviously keeping on track of this one. It will be good to keep in touch with him and see where this goes and how it develops. He seems determined to pursue it across the world."

"Yes, he does seem determined. Thank you, Ali." Joe put the phone down and by the time the phone reached Joe's desk, the decision was made.

Ali held an issue of the magazine, which had arrived that morning by mail. He checked the masthead, as he always did, and, as always, his name was there as consultant. He was getting used to seeing his name in print, as he had published a series of articles about his experience as a surgeon and consultant. Joe had been very helpful in honing his writing skills, giving advice and skillfully editing his produce, so that now he felt confident enough to be writing a book on advanced surgical techniques. He reread Jean Legrande's article and noticed that separate notation had been added, describing two further cases—one in Tennessee and one in Florida. Fertility was not exactly his specialty and he was only interested in the subject as far as it might have some widespread repercussions, influencing the practice of medicine in general.

Ali considered himself a very modern practitioner, and although he was a surgeon, he strongly believed that prophylactics were far more important than a cure. And even as far as healing is concerned, he regarded surgery as a last resort. In his consultations he was always looking for alternative methods, to avoid, or at least delay, the day when a patient would need to be subjected to the knife, or scalpel. He was known for his steady hand and reasonable approach, and there were always far more patients seeking his advice than he had time to see.

Ali could see that the trend in the world over the past decade had been an increase in diseases, and he instinctively knew that this trend would continue until many more humans learned to live in harmony with the natural worlds. He also knew that it would be impossible to go back to living in huts and to become an agrarian society again, so he frequently pondered this dilemma, wondering what the future of his children's children would be like.

When Ali was young he experienced an electric shock when one of his visiting aunts shook hands with him when he was only six years old. This aunt was known to be a healer and when he quickly pulled his hand away from her grip, the aunt looked at him quizzically for what seemed to be a long time. She then explained that he had felt what she called "the healing force." Ever since then Ali became interested in the electricity and power of the human body. He soon learned that all organs and all cells thrived on subtle electric signals and that energy—that unseen but ever-present power—was as much part of the healing process as any herb, potion or pill.

Joe arrived home tired. He got tired so much faster these days. After all, he was older and his responsibility was much greater. He briefly thought back to those carefree days when, as junior editor he would be sent on an assignment and his main preoccupation of the week would be to write his article or report in time for the editorial deadline and to do his best so that he could successfully advance his career. In those days he longed for more responsibility and now that he had it, he wondered whether it had been worth it. He also noticed that he had put on some weight and that he tended to spend more time reminiscing about his life's journey than planning ahead.

As he drove into the driveway and stopped the car

## The Race Toward Success

in front of the house, he paused for a moment to breathe in the fresh, suburban air, before going into the house. Inside the house was Maggie—his wife and mother of their sixteen year old daughter. Nine years younger than him, she was still good looking, with that rich auburn hair of hers, her lovely smile and sunny disposition. How was it possible that this was his life? How come a woman like Maggie, who could have had any man, had chosen him above all others? He thought back to their first date and how he was smitten by her, as he assumed all the other men on the staff of the magazine were. He remembered walking by her desk as often as he could make it seem reasonable, sometimes stopping to have a brief word, but not daring to be too personal or to ask her for a date. And then that wondrous, surprising day when she made it clear to him that she would not say no to a date. How did she do that? He tried to remember. Oh yes. It was a lovely spring day and she had been part of the staff for about six months, typing up their articles, answering the phone, opening up the mail, running to the bank and generally keeping them all in order, reminding them all of their deadlines and delivering the completed issue to the printer via e-mail, and arranging delivery to the distributor. They all loved her warm smile and benefitted from having a female amongst them. She was not part of the team, and yet she was an integral part of the monthly process of producing a magazine.

    A famous doctor, a Nobel prize winner was in New York for a seminar, and Joe was asked by the chief editor to interview him. Joe was very proud of the assignment, though he realized that it was only because Bob Harris, whom he was assisting at the time, couldn't make it. He was visiting a new cancer research center on the west coast and would be gone for a couple of days. So it was up to Joe to go to the seminar and interview the eminent doctor. The subject of the seminar was *The Prospects of Genetic Engineering*, about which he knew very little, but that did not matter. He would go to the library and read up about it. After all, the doctor was the specialist, so he would provide all the relevant information; Joe just needed to ask the right questions. He had already conducted a number of interviews for the magazine and he had found out that it wasn't that difficult—the interviewee was usually more nervous than he was.

As he was leaving the office that afternoon to go to the library, he walked by Maggie's desk on the way out. She was getting ready to go as well, so he instinctively stopped and waited to walk out of the office and down the stairs with her.

"I'm going to interview Doctor Stanford," he boasted, just to make conversation.

"Oh, that's wonderful, Joe" She seemed to be genuinely pleased for him.

"Yes, but I have to sit through an entire day at the seminar, listening to him, so I know what his work is about."

"Well, that might not be so bad; in fact it can be very interesting," she reassured him. They had just reached the ground floor and had paused in front of the door leading onto the street. Joe looked at Maggie and she smiled at him encouragingly, and it was at that moment that he decided to take his courage in both hands and ask her to join him for the cocktail party that was to be held after th seminar. "There is a reception after the seminar at the Grand Hotel," he said, a little too fast. "Would you like to join me there as my guest? They've given me an invitation for two. At least I will know someone there, if you come."

"Yes, I'd love to," she said, totally unexpectedly. He could feel his heart racing.

"All right then, meet me in the lobby of the hotel at eight tomorrow evening and I'll come and fetch you."

"Okay, see you then."

He walked in the opposite direction, going left away from the office building, even though he usually walked to the bus stop, a couple of blocks away to the right. But he didn't know what to say and he worried that she would notice him blushing and stammering and that she might change her mind.

The next day he found it difficult to concentrate. Nevertheless, what he did hear was fascinating. He had no idea what experiments were being conducted with genes—mixing animals with plants and trees, creating endless new varieties. He wondered whether vegetarians knew they were consuming animal genes when they were eating tomatoes with penguin genes to make them resistant to the cold, or apples with elephant genes, to make them big. There still wasn't the legislation in place to make these mutations clearly marked on the supermarket shelves. It made him wonder how far human genetic engineering had advanced. He also wondered whether if anyone got hold of and exposed the truth, they would make a lasting contribution to public knowledge concerning contemporary medicine and its practices. He thought maybe this was an undertaking he could initiate, and he wondered how realistic his chances of doing so were.

Doctor Stanford turned out to be a good speaker. He was very entertaining, describing with vision and humor the future of genetic engineering. He had good slides and film clips showing several interesting mutations, as well as genetically engineered cures. He believed totally in what he was doing and promoting; he was working on defining deficient genes to help find cures for all major diseases—from cancer to MS, Parkinson's and AIDS. The seminar itself was an attempt to raise funds so that he would be able to continue his research and establish a new center for genetic engineering with modern laboratories and permanent research staff.

Joe had bought some quality notepaper and matching envelopes in a gift box; he also purchased a couple of good pens and some stamps, which he added to the box. He wrote a small card and put Mario's name on the envelope. Something had caught his imagination about this young man who was dying. It seemed very unfair to Joe that someone as young, with so much to live for, would need to leave this life with no support from those who were nearest and dearest to him.

As he walked into the ward, he spotted Mario, sitting up in bed, eating his lunch. He looked a little bit better and Joe wondered whether the medicines were working, after all.

"Hi, Mario. How are you doing?" he asked, trying to sound as cheerful as he could.

"Not too bad, not too bad," came the response. "I have good days and bad days. Today is one of my better days."

"Well, that's great," Joe said and then he held out his hand with the box in it. "I brought you something," he said.

Mario's face lit up. "Really?" he asked, incredulously. "That's for me?" He looked almost as wide eyed as a child on Christmas morning.

"Sure," Joe replied. "Open it." Mario opened the box and gazed at its contents for the longest moment. "These are beautiful," he said, in tones that were barely audible. "Thank you so much." He reached for his bedside table, but Joe was quick to stop him.

"No, please. This is a gift from me," he said.

"Well thank you, Joe. That is so kind of you."

"And if you want me to mail any of your letters, I would be glad to do so," Joe said, realizing that he had just committed himself to a return visit. He drew up a chair close to Mario's bedside. He had brought with him the new edition of the *New Medical Research Magazine* to show Mario that his story had appeared in print. Mario looked at his picture and glanced over the article, but he didn't seem keen on reading it at the present time. He placed it on his bedside table and asked, "Can I keep it?"

"Yes, of course, it's yours," Joe replied.

"You know I used to be a violin player with the Harmony Quartet. Have you heard of them? Or perhaps you have heard them playing?"

"No, but I have heard of them. I've heard they are very good."

"Oh, yes!" Mario's face lit up. "We were really good. We used to travel around the country. Every month a different state. I've driven up and down the country so many times. It was great." He closed his eyes for a moment. "We played all the classics—Vivaldi, Boccherini, Beethoven, Bach, Mozart, Ravel—you name it, we played it. Some modern pieces too. Critics used to say that our interpretations of the classics were aggressive. People either loved us or hated us. But we did fill the audiences."

"Well, I wish I had seen you play," Joe interjected and Mario seemed to return to the present, rather than reliving his days of glory.

"I don't think you are going to hear me now," he said with regret. "I don't think I am going to get out of here alive." Joe wanted to contradict him, but he felt it would be dishonest to do so, so he kept silent.

"Do you believe in reincarnation?" he asked, thinking that if Mario did, it would offer him some solace.

"I don't know," the young man replied. "I do believe there is something, though. Can I tell you about it in private, I mean off the record; just between you and me?"

"Yes, of course you can."

"Well, last night I felt I was floating above my body and in fact I could see myself lying beneath me on the bed. And then I noticed someone I had never seen before floating beside me. He was waving to me, as if he was encouraging me to come with him, but I resisted. I thought it was a dream, but today I saw them wheeling someone down the corridor as I was walking to the toilet. I was curious, so I looked, and it was the man who had floated beside me last night. He looked dead—he was pale as a sheet and he didn't seem to be breathing. What do you make of that?"

"I don't know," said Joe. "It sounds real, though. I have heard of what are called out of body experiences, where you float out of your body. It sounds like it was one of those—an OOBE." He laughed at the acronym.

"Putting a name to it doesn't explain it," said Mario in a matter-of-fact way.

"I guess not," admitted Joe.

"Anyway, I don't think I could tell that story to the doctors here or to the nurses. Even my sister would probably laugh at me."

"Don't worry, your secret is safe with me."

At lunchtime there was an hour's break in the seminar proceedings and Joe, left to his own devices and wondering where to go for lunch, realized that the hotel was only two blocks away from the hospital. In fact, he had passed the hospital on the way in to the hotel in the morning and had thought of Mario, feeling guilty that he had not been to see him for several weeks.

"I'll go and see him now," he thought as he walked through the revolving door of the hotel. He walked fast, determined not to be distracted on the way. When he arrived at the hospital, he went straight to the elevator and pressed the button for the third floor. Then from there he turned right, past the nurses' station and onto the AIDS patients' ward. The large room was brightly lit by sunlight coming through the windows and Joe paused in the doorway, trying to locate Mario's bed. He remembered it was the third bed to the left, but that bed was empty, neatly made up, with no flowers or personal possessions on the bedside table. He then started to look at the other patients, slowly walking to the end of the ward and then turning back toward the door. There were several men occupying most of the beds, some with visitors sitting next to them, others barely conscious, and a couple of others watching television suspended from the ceiling above their beds. Mario was nowhere to be found.

"They must have moved him," thought Joe, as he walked back out of the door and paused at the nurses' station to find out.

"I'm looking for Mario Calioni," he said to the nurse behind the desk. "Has he been moved from the AIDS ward?" he asked.

"Mario died this morning," she said slowly, not being sure who Joe was to Mario.

"Oh, I am sorry," Joe said, as a wave of guilt hit him. He had nothing more to say, so with a brief "Thank you," he turned away and walked toward the elevator.

On the way back to the hotel and the seminar he sat down on a bench and spent a few minutes in contemplation. He felt bad that he had put off his planned visit to see Mario. He also considered his own mortality, as he wondered where Mario was and whether Mario could see him now, and whether he understood or felt his grief. It was quite intense and Joe sat there for a while, oblivious to the time and place. He was late for the next session of the seminar and throughout the three hours he found it difficult to concentrate on what was going on inside the auditorium, with his mind racing back to the hospital and wondering about Mario's final moments. If only he had gone in yesterday, or even this morning, he would have still seen Mario alive!

When the conference finished, Joe was the first in line to speak to the professor. He had plenty of notes; enough, in fact, to write a whole article about genetic engineering, even though in the morning he still knew very little about it. He had photographs from the conference press office, facts, figures, examples and... worries. He knew that the controversy surrounding the morality of genetic engineering had been going on for a long time, and that there were not only those who opposed it, but also those who claimed there were diseases emerging because of it. Joe felt ambivalent about the whole thing and wanted to present as unbiased a view as he possibly could.

One thing Joe wanted to ask about was human genetic engineering. So he held up the microphone and asked the professor if he could interview him for a report from the seminar he was writing for the *New Medical Research Magazine*. The magazine was beginning to gain a favorable reputation among medical professionals and Joe was pleased to hear the great professor's response. "Let me just speak to these people," the professor indicated the line of people forming up behind Joe. "And then I'll see you in the seminar admin suite on the first floor. Room 112. I'll be there in fifteen minutes."

Joe wandered around the auditorium waiting for the professor to finish answering individual questions. He looked through the dozen or so books the professor had written, as well as other literature advertising further seminars and events. This was one prolific man with varied interests. It seemed the professor was concerned about the future of the human race and was predicting the appearance of new forms of diseases. He also seemed to be very optimistic about the potential of not only genetic engineering, but the ability of medical scientists to clone any organ, any part of the human body, by altering the genetic coding of individual cells. Joe began to have visions of human parts factories and perhaps organ mail order warehouses. He imagined a catalog of parts and smiled to himself as he thought how ridiculous that was. Nevertheless, Hollywood was already exploiting the subject with films about cloning experiments gone wrong and exploring the morality of creating new humans.

By the time Joe looked toward the front of the auditorium, the professor had gone and there were very few people left in the room. He dashed downstairs and quickly found room 112. He knocked on the door and, not hearing a reply, he turned the old fashioned handle and entered what turned out to be a suite. The room he entered was set up like an office with three desks. Behind two of them sat a man and a woman, with the woman sitting closer to the door. The man was on the phone discussing arrangements for a further seminar abroad, whilst the woman was typing something into a computer. Joe walked up to her and said that he had come to see the professor.

"Go right in," she indicated a door next to her desk. Joe walked up to the door and knocked.

"Come in," he heard from the room behind the door. Joe turned the handle and walked into a large living room with a couple of sofas besides a fireplace, a few armchairs, a writing desk and a table with chairs around it. There were cut flowers on the table and on the mantle piece. The professor was sitting in front of the fireplace.

"Come in and sit down," he said. "You wanted to interview me for the *New Medical Research Magazine*."

"That's right," Joe replied. He sat down on the sofa and took his tape recorder out of its bag. As he was connecting up the microphone, he added, "I just want to ask you a couple of questions."

"Okay, shoot."

Joe turned on the recording machine and asked, "I want to know if it is true that genetically engineered humans already exist."

The professor smiled indulgently. "Well, that is such a broad term. Do you mean, are there human beings that have been genetically altered in any way?"

"Yes, well, let's start there," Joe replied.

"Of course there are. Probably thousands. Maybe even millions. You don't hear about them in the papers, but human genetic research and experimentation are very advanced in many countries. They are not publicized because of public opinion and the hue and cry that would be raised against those who are conducting the experiments. It would take years of legal wrangling and appeals to be able to officially even begin what has been going on in laboratories all over the world for years." Joe was surprised. He had had no idea, but was determined not to show his ignorance.

"What about genetic mutations? Have humans been involved in some of the experiments, like the ones you mentioned in your seminar concerning vegetables and animals?"

"Oh, yes, of course. It goes on all the time and researchers are learning something new about what works and what doesn't work every day."

Joe felt sick. An image flashed through his mind of a tomato with human eyes, looking at him, following him around. Certainly that would be an experiment gone wrong.

"Can you be more specific, Professor?" he asked.

"No, I cannot. I don't want to get me, you or your magazine into trouble. I don't know the details anyway. It's not regularized and it's not legal. So let's leave it there, young man."

Joe agreed reluctantly and asked a few further questions concerning the professor's background, his plans for the future, and the seminar itself. When he thought he had enough material to complete his article, he thanked the professor and got up to leave. As he was packing up the microphone and the recording machine, the professor gave him a curious look and said, "Off the record, do you know what my next book is about? It might bring me a second Nobel prize. It's so much easier to win a second Nobel prize than a first."

Joe did not understand. "Why is that?" he asked.

"Because to get the first one, you have only one chance in millions, or even billions; to get the second, your chances increase to one in five hundred, which is the number of Nobel prize winners alive today."

"So what is this new book about?" Joe asked, as he closed the case that housed his tape recorder and microphone. The professor stood up from his armchair, took a couple of steps closer to Joe, so he could whisper and still be heard. "It explains how genetic change happens in the presence of ultra-violet light. Humans are natural manufacturers and transmitters of ultra-violet light."

"I see," Joe said as he picked up the case, knowing that he did not see at all.

"I don't think you do," the professor said and Joe stood still, waiting for an explanation. "Do you know that monarch butterflies, when they migrate to their breeding grounds in Mexico, they breed there and die?" Joe nodded and the professor continued, "The next generation somehow know where to go and during the next season they return to the exact same place where their parents started from during the previous fall. How do they know?"

Joe shrugged his shoulders, indicating that he had no idea. "I don't know," he said. "How?"

"It's in the genes. They are born with the knowledge where they need to go and generation after generation, year after year, they follow that print." The professor was getting excited and his gestures were becoming more animated. "We humans have the same capability. We can pass on information from generation to generation. The difference is that mostly we no longer know how to decode the messages that are passed down to us in our genes from our ancestors. But the information is there and I am working on ways to uncover it and utilize it to ensure survival of the species."

"A noble undertaking," Joe said. There was silence for a moment, and then Joe added, "I hope your tour goes well," and they shook hands, as if there was now some kind of conspiracy between Joe and the professor.

As Joe walked downstairs to the lobby, he was still deep in thought, with images of mixed species swirling around his head. He was also still under the influence of grief, not having been able to see Mario alive again before he had passed away. At the bottom of the stairs he looked up and there in front of him, walking through the door was Maggie, looking radiant and beautiful in a bright blue dress and a black jacket. She was wearing black high heels and had a black shoulder bag with a golden chain over her right shoulder. Her hair was piled up on top of her head with escaped strands of auburn hair framing her young face.

"Maggie," he exclaimed as he walked toward her, "I am so glad you were able to make it." They stood together in the lobby for a moment as Joe asked Maggie whether she wanted to go straight into the reception room or whether she would want something to eat first.

"I've had dinner," replied Maggie, "so we can go in," she replied, but just as he was about to turn toward the elevator and guide her onto the fifth floor which is where the reception was being held, Maggie hesitated and said, "Wait a minute. Tell me what's wrong. Something is wrong, isn't it?"

Joe looked at her and wondered how she could possibly tell.

"Yes, it is. A friend of mine died today," he replied. Maggie sighed, not knowing what to say.

"He died of AIDS in the hospital," Joe added by way of an explanation.

"I am so sorry," she said and she really looked sorry.

In fact Joe began to regret the fact that he had confided in her.

Maggie spoke again, "Perhaps you want to talk. Or maybe you would rather be on your own," she volunteered. "This might not be the right time for a cocktail party."

"No, that's okay. Let's go. It might be the best thing to do right now. I don't know any more. I think he would have wanted me to go."

The elevator arrived with a ding and they entered, accompanied by a number of other patrons and well dressed people who were arriving for the party. They pressed the button for the fifth floor and and the elevator glided upwards with everyone watching the illuminated numbers above the door shifting from number to number.

As they arrived, and slowly walked toward the entrance to the large reception room which had a bar at one end and a stage at the other, they were greeted by Professor Stanford, accompanied by the two organizers of the tour, whom Joe had seen earlier.

"Welcome," the professor extended his hand. "I look forward to seeing that article. And I hope you remember what I had told you."

"I don't need to," replied Joe "I have a recording. And by the way, may I introduce Maggie Holland?"

"Delighted to meet you," replied the professor as he shook Maggie's hand. He then turned back to Joe. "I meant what I had said after the microphone was off and you had put it away."

"Oh, I thought that was off the record," Joe said. "I wasn't going to mention it in my article."

"Oh, you can, by all means, as long as you do

mention that it was off the record."

"I see. Thank you, Professor," Joe seemed confused, but decided not to ask any more questions, as there were others lining up behind him. "Is it off the record, or isn't it?" he wondered, but was satisfied that he could use the information he was given.

Jean Legrande picked up the phone. He was back in Paris, preparing for an interview for *Paris Match* when the phone rang. It was Joe. At last! He was beginning to wonder whether he should sell his article to another publication.

"Jean, it's Joe. We've decided to publish your article," he said.

"Oh, good. There is a lot of interest here. I mentioned it to a few of my friends in the publishing business, and they think I should write a book."

"Well, that sounds like a good idea. In the meantime this should give you some publicity over here, so that after your book is published, you might want to do a lecture tour."

"Would you organize it?"

"I don't know. I'd have to think about it." Images of the hotel suite all those years ago with a secretary and public relations manager flashed through Joe's mind. Professor Stanford had passed away, but his work was still influential in the medical world. Perhaps Joe could put something together. But in the meantime he needed to ensure that copy was ready for the next issue.

"Got to go. Just wanted you to know that it will be coming out in a few weeks."

"Thank you for telling me."

Jean was thrilled. At last, things were going his way. He had managed to publish first, right in front of Elizabeth. Well, that wasn't fair. She had, after all, done some of the work, but not as much as Jean had. He had traveled extensively on two continents to gather samples and to write his conclusions down in the form of a scientific hypothesis. He had, after all, mentioned Elizabeth and he was certain that she would be grateful for this generous inclusion in his groundbreaking discoveries. He had enjoyed his session with her and he would remember her if anything new came up, but one has to give credit where credit is due and omit credits where they had not been earned.

# CHAPTER SIX

# A Disturbing Trend

It was nine o clock in the morning as Anthony Pontrick kissed his wife goodbye, and headed out the door on his way to work. As always, the chauffeur driven government subsidized black limousine was waiting in the driveway with the engine running, facing the road, ready to depart. As Flavia opened the door for him, her eyes fell on the semi-circular hall table by the front door, where the morning paper was resting, having been picked up from the front porch by their daughter Mary, as she left for school, just in time to catch the school bus as it stopped in front of their suburban house.

"Don't forget your paper!" she exclaimed as she reached for the *Morning Post* with her left hand, still holding the door with her right. He paused in the doorway as he turned back toward her, waiting for his morning supply of news, views and gossip.

"Thanks, honey," he said, smiling at her, as she gave him the paper and kissed him gently goodbye.

"Don't be late tonight," she reminded him as he stepped onto the porch and she held the door, calling after him. "Remember the Bunters are coming to dinner."

# The Tale of the Tailless Sperm

He didn't turn around, but waved his right arm with the newspaper in it, so she could see that he had heard her, and that he had remembered and would do his best to be back on time. He was already on his way; his mind was racing ahead to the meeting he needed to attend that morning and all the other chores that lay ahead of him.

He was a tall, thin man in his late forties, and his long, black, camel coat made him look even thinner than he actually was. He had a long pointed nose and a thin, intense mouth, but he still had a full head of black hair, only slightly greying at the temples, with a few more strands of silver shining through. He exuded energy and decisiveness as well as a certain intense preoccupation, even when he was not preoccupied, which was very rare indeed. A man with a certain charisma and leadership quality, who believed in what he did, Anthony had received the best education the country could provide. He had majored in philosophy and political sciences, graduating with honors and beginning his career as a junior assistant to the head of department at the Ministry of Health and Welfare, which he was now running. Twenty years had passed and he had established himself as a keen policy maker and manager, a man with vision and a sensitive sense of the need of the times. He was often called in to conferences with the minister and his deputy, when important policies were being discussed and decided. There was talk in the ministry that he could be the next minister or at least his deputy, though it wasn't certain whether Anthony's politics would allow for him to travel to the top of the ladder. He seemed to be more interested in policies than politics and more concerned with the effectiveness of his work than whether his superiors were satisfied with his performance. With a reputation of being his own man and having a history of rebellious behavior behind him, Anthony would probably remain where he was, programming future research trends and developments, as well as government sponsored programs. He did have at his disposal a considerable annual budget which he implemented thoughtfully and frugally, saving the government millions each year, but putting the money he did spend to good use.

On this Tuesday, as he got into the car, he relaxed into the leather of the back seat of his government chauffeur driven limousine—one luxury he really did enjoy. This was his time, a time to relish his morning paper so that he could become updated about world and local news. He was proud of the fact that he was well educated and well read, and reading the morning paper was part of his daily ritual, a habit of many years. He first read the headlines, then scanned the entire paper to see if there was anything of interest on the remaining pages, finally returning to those articles which interested him the most. By the time he would arrive at the ministry, he would have a good idea of what was printed in the *Post* that morning, even if he did not have time to read all the articles that he would have wanted to. There would always be time for that later, when he had his tea break later in the morning. Anthony liked his routine and tried not to veer too far away from it, for it gave him a sense of stability, satisfaction and a job well done.

So this day the headlines, as usual, were disseminating information about disasters and wars—

there was an earthquake off the coast of Japan and two hundred people living on a small fishing island were rendered homeless. There was civil war in Africa and the police had managed to break a large drug ring in England. Then there were adverts, theater and film reviews, social commentaries and news about stocks and bonds. And then on page 16 a short article caught Anthony's eye—a brief reprint from a medical journal mentioning a new form of disease that was causing impotence in men. Just a brief note, but Anthony sat up straight in the back seat of the car, as he memorized the title of the journal mentioned. He recognized the name, as it was one of the several magazines he had his secretary subscribe to, because he felt he needed to know what was going on within the discipline he was responsible for. So any new medical discoveries and inventions would find their way to his desk, whether through the official channels in reports and applications for government funding or patents, in proposals by firms and individuals, or in articles and reviews from medical literature. For a while now he had noticed a trend in the diminishing birth rate among families in developed countries and wondered what was the cause. As he arrived at the ministry, the doorman walked up to the car and opened the door. Anthony stepped out.

"Good morning, Gaston," he said, as he always did. Gaston wasn't French, but originally his parents came from Jamaica. He was very black, with a great big smile that always made Anthony feel both important and welcome.

"Good morning, Sir," Gaston replied. "And a

beautiful morning it is, too." Anthony hadn't noticed it before, but now that Gaston mentioned it, he suddenly felt the warmth of the sun penetrating through his coat and warming his back.

"That it is, Gaston; that it is." He walked past the doorman and headed for the elevator that would take him up to the seventh floor where his office and department were located. Once on the seventh floor, he felt at home, for this was his domain. He entered his office and nodded to his secretary as he walked past her desk.

"Good morning Mr. Pontrick," she said, as she immediately got up to make him his morning cup of tea. He was not a coffee drinker, but liked his tea strong and served with two lumps of sugar (lumps, not spoonfuls). This was their morning ritual and they both knew it and abided by it. He would have his tea while he perused the professional magazines and reports, and she would hold his calls till after 10, if she possibly could. Of course, sometimes he was summoned onto the twelfth floor to meet with the minister and perhaps some other distinguished guests, and then the routine would be broken. But otherwise it was adhered to, if at all possible.

Anthony entered his "den" as he liked to call it, threw the paper and his briefcase on his desk and hung up his coat in the closet by the door. He then sat down in his large leather executive chair and began to look through the pile of papers, periodicals and reports that awaited him on his desk.

"Now where is it?" he muttered to himself, looking for the *New Medical Research Magazine*. There was a knock on the door. "Come in," he called out, not looking

# The Tale of the Tailless Sperm

up from his tasking. It was his secretary, Mrs. Green, with the tea. Ah, it was so good that some things were still reliable and very pleasing to the senses. He leaned back and smiled at her. "Have you seen the *New Medical Research Magazine*, Mrs Green?" he asked.

She put the china cup and saucer with a blue design on it down on the desk. The two lumps of sugar were resting on the saucer, together with a silver plated spoon.

"It should be there," she said. "In fact I am sure it is there. I put it on your desk yesterday afternoon." She thumbed through the pile and, remembering that it had a bright blue cover, she quickly found it. "There you are Mr. Pontrick," she said as she handed him the magazine and turned to the exit. He took the magazine from her, sat back in his black leather executive chair and began to flick through the pages, until he found the article he was looking for. The title jumped out of the page: *A Disturbing Trend*, and there were photographs which he immediately recognized were enlargements of micro-photographs of sperm. He wasn't sure at first, because they were not normal, but they were photographed gathered around a giant egg—he knew the proportions of the two (one was many times larger than the other). Something, though, was terribly wrong because the usual long, vibrating tail of the sperm, which enabled propulsion, was simply not there. The picture was an artist's interpretation, and clearly the sperm were not being very successful at impregnating the egg, even though it was right there and they did not need to make the journey through the uterus up the fallopian tube to get there.

Anthony picked up a pair of scissors which he kept

in a pencil and pen container on his desk and cut out the article.

He opened the bottom drawer of the desk, which was set up as a filing cabinet, and pulled out a file marked, "Trends to observe," and filed the article away, making a mental note to keep an eye on this one. He took a label from the top drawer of his desk and wrote on it "Legrande Syndrome" and pasted it on the inside flap of the file. There were a few other labels already attached to the file, all with names of new diseases, or trends, or drugs from abroad that he would look up on the Internet or enquire about at a later date, when he felt there was enough momentum attached to the subject to offer further intelligence to work with. He liked to think of himself as a man who kept his finger on the pulse of current trends, so that he could then approach his minister or vice-minister with updates, suggestions and proposals. In fact, he believed that he had come this far in his career precisely because of this aspect of his mental agility and initiative. It was like having a sixth sense and being able to spot trends during their incubation period, so by the time they became widespread or dangerous, he would be ready with the relevant research and prophylactic suggestions. Like the time he suggested the use of temperature measuring machines for screening airport passengers for SARS at a distance. Not very popular with the passengers who might have only had the flu, but effective nevertheless. SARS had been contained in its infancy and was no longer a threat; more like a bomb that had never exploded is the way he liked to think about it. COVID-19 was, of course, another story, and he had very little to do with its demise.

# The Tale of the Tailless Sperm

It was a vaccine which in the end conquered the virus, and he was sorry to admit that it was the Brits who found the effective and safe prophylactic, despite the efforts of his friends and colleagues at the major drug companies.

# CHAPTER SEVEN
# The Plot Thickens

Jenny sat opposite her friend Jill. Jill taught at the same school as Jenny and was responsible for the sports and gymnastics activities. Every day she would take Jenny's class for an hour of physical exercises, either in the park or in the gym. This allowed Jenny to have a lunch break and a breather before the afternoon lessons. Jill was Jenny's age and single, and she had very definite views on just about everything.

They were sitting at an outdoor table at a French style bistro located close to the school.

"So why do you have to have a baby?" demanded Jill. "There are too many children in the world as it is. Why don't you adopt an orphan who badly needs a home?"

"Oh Jill, you don't understand, do you? I want a baby of my very own. One that will look like me and Dick, and carry our genes. I've heard so many stories about people who adopt children and then don't get on with them at all. Or they could be carriers of congenital diseases. Who knows what you're getting when you adopt a child?"

"No, you don't understand," said Jill, stressing the

word *you*. There are agencies who specialize in checking these things out. You can do the appropriate research and have a pretty good idea about a child's background. And you would be helping the world situation. On the other hand, if you have your own baby, you still don't know how he or she will turn out. What if you have a baby the natural way and he or she turns out to be a thief or a criminal?"

"Well, maybe you're right. But the bottom line is that we want a baby of our own, so we're going to do everything we can do to have one."

So that was the end of that and they both realized they would never discuss the adoption option again. There was an uncomfortable silence between the two friends as they drank their coffee and avoided looking at each other.

At last Jill broke the silence, "So how is Dick doing?" she asked.

"Oh, he's fine. I think he is beginning to get impatient with me, going to all these doctors."

"Well, doesn't he want a baby, too?"

"Yes, he does, but mostly because it will make me happy. I don't think he minds too much one way or the other."

"Well, say hello to him from me. I've got to go."

Jill gathered her purse and sunglasses and got up. "I'll see you Monday," she said and headed toward the door.

The Minister of Health and Welfare was fat. Anthony had no respect for this man who looked ill and was certainly not a fitting advertisement or representative for the domain he lorded over. But he was a party man and Anthony knew that he would go just as suddenly as he appeared in the office suite upstairs. Anthony also knew who was really boss and when dealing with Mr. Sedwick, he adopted a somewhat indulgent and patient air, as if he were dealing with a child.

As he knocked on the door, he prepared himself for a grueling session. It usually demanded from him a large dose of tact and skill, to educate his minister about the latest developments in health or medicine, while playing the game and pretending to be instructed by his boss, adopting at the same time a submissive attitude. Needless to say, this attitude he tended to adopt was both difficult and tiring, but one that he deemed necessary, if he wanted to keep his job.

When Anthony entered the minister's large office, he noticed with relief that Mr. Sedwick was smiling. He was sitting behind his desk with his large frame completely filling out his executive chair.

The minister waved his hand. "Come in, Anthony," he said. "I need to talk to you."

Anthony walked up to the chair located in front of the large mahogany desk. He briefly remembered how this desk was specially imported from Sweden to give Mr. Sedwick the kudos and sense of importance his predecessor obviously did not enjoy, as he had sat for years behind a government recommended L-shaped computer desk. A brief smile appeared on Anthony's lips, soon to disappear as he focused his attention on what his boss was going to say to him next.

"I have been pressured by the lobbyists," Mr. Sedwick said. "We must use our influence to help support the pharmaceutical companies. They are losing money to those alternative healing methods. Perhaps we have made a mistake by making them legal and allowing them to continue."

"It seems to be too late to introduce legislation that would take away their freedom to practice," said Anthony.

"Ah, that would be nice." Mr. Sedwick briefly closed his eyes as if imagining a world without acupuncture, massage, Reiki and Therapeutic Touch. The image seemed to give him great pleasure because he smiled again briefly. Then suddenly he opened his eyes and snapped back to reality.

"Our policy and our campaign was based on the individual's freedom of choice," he said. "This applies to a person's right to choose their insurance policy as well as their healing modality. We can't take that back, though no doubt it would benefit the population at large. No, that is no longer an option," he confirmed. "But we can help promote old-fashioned cures, like pills and potions. Especially if we can find a new scare and promote legislation inforcing immunization or some kind of prophilactics. Is there anything like that you can find?"

For a moment Anthony found himself thinking that the real old-fashioned remedies were in fact such cures as acupuncture, herbs and massage, not pills or jabs.

"We're past the flu season," he said. "Diseases are not on the rise at the moment," Anthony added. But I will keep my eyes open for new diseases or syndromes." He was getting ready to get up and leave.

"Do hurry and find something soon," Mr. Sedwick said. "We badly need those contributions from the pharmaceutical companies."

Having said the word "syndromes," Anthony suddenly thought of something. He quickly lowered himself back down into the chair he had just vacated.

"There is something…" he began tentatively.

Mr. Sedwick leaned forward, resting his large forearms on his enormous desk.

"Yes? What is it?" he asked, exuding more energy than Anthony had ever witnessed before.

"It's called The Legrande Syndrome," Anthony replied.

"Never heard of it," Mr. Sedwick said.

"That's because it's quite new. I've just found out about it today myself."

"Well, what is it?"

"It is a strange phenomenon," Anthony explained. "It appears that all over the world there are men who are producing sperm that lack a tail and are not able to fertilize an egg." He realized that he was embellishing the story somewhat and that it was still too early to say for certain whether the tailless sperm were able to penetrate an egg's defences or not. Also, "men all over the world" so far amounted to fewer than twenty, but it certainly sounded like it could develop into the beginning of an epidemic, or even a pandemic.

Mr. Sedwick seemed very interested. "Is there a preventative measure we can recommend?" he asked. "Something that the pharmaceutical industry could produce cheaply and in large quantities in a short space of time?"

"I don't know yet, but I'll find out," Anthony said, wondering at the same time whether he would be able to easily communicate with this Legrande fellow. He seemed to remember that the good doctor lived in Paris.

"Then find out," Mr. Sedwick said. "I want more information as soon as you can get it."

"Yes boss," Anthony said, standing up. He then walked toward the door of the office.

Back in his office, Anthony searched through the pile of newspapers and magazines on his desk.

"Where is it?" he muttered to himself as he rummaged through the stack of paper for the second time on that day. Although his secretary had identified the magazine earlier that day, it had been covered up by someone delivering a stack of magazines he had received while being in conference with his boss.

"Where is the *New Medical Research Magazine*? Where is that article?" He soon found it—the magazine was left open on the page he was searching for.

"What was her name? Elizabeth something." She had found a couple of cases in her practice as OBGYN. He scanned the article and found Elizabeth's name mentioned toward the end. In fact, it was in the last paragraph that Legrande had mentioned her contribution to the discovery of The Legrande Syndrome. "That's right—Elizabeth McNeill. That's her name." Anthony Googled her name and soon found the website of the clinic she had founded.

# The Plot Thickens

He dialed the number and was put through to Elizabeth's extension—the doctor was in.

"My name is Anthony Pontwick and I am a director at the Ministry of Health and Welfare."

"What can I do for you?" asked Elizabeth. She did not sound very encouraging.

"I have read Jean Legrande's article and we are getting concerned that this could become a worldwide pandemic. Have you had any success with these tailless sperm? Have they been known to fertilize an ovum?"

"Not yet. I have placed them in a culture spiked with hormones, in close proximity to an egg, but they seem incapable of a response that their tailed counterparts are guaranteed to perform. They just wriggle and go nowhere."

"Have you tried giving them a little push?" asked Anthony, smiling to himself as he imagined the doctor reaching into the solution with some instrument and attempting to guide the reluctant sperm toward an ovum. He realized that his question must appear to Elizabeth to be very non-scientifically worded.

"Oh yes," replied Elizabeth, who understood the question. "I have tried everything. I have even injected the little buggers into the egg, but nothing happened. The egg completely ignored the fact of being impregnated and did not divide. It just absorbed the sperm which consequently disappeared, as if nothing had ever happened. You do realize that the sperm is the smallest cell in the body, while the egg is the largest. It is like you or I walking into a sky scraper. It's easy to disappear without a trace. I am still working on it," she added. "It's not an epidemic yet. Two cases here and a couple more in Tennessee or Miami do not a national disaster make."

Elizabeth was surprised the government was getting involved at all.

"Keep me informed," said Anthony and gave Elizabeth his office number as well as the number of his cell. He then put the phone down.

The next call Anthony made was to Andrew Stewart, the CEO of Visor Pharmaceuticals. Andrew was in his fifties, a man who originally inherited a place on the board of the company. His father was one of four sons who had inherited the company from their father, who had started off as an owner of an apothecary in the 1890s. He had made his fortune by importing a number of exotic herbs and mushrooms from the Amazonian jungle after his years of religious dedication as a missionary. When he came back to the United States he started producing a number of drugs that turned out to be beneficial in combatting such common complaints and diseases as whooping cough, diphtheria and the common flu.

He then branched out into nuts, supplements, and foods, adding exotic berries and fruits to such snacks as crackers, cookies and cereals. His greatest success came when he bought a large farm covering several hundred acres and started growing his newly discovered vegetables and fruits closer to home. He settled down and married the daughter of a distillery owner who was the only heiress to her parents' fortune. Their wedding took place in 1910, twenty years before prohibition when the distillery was forced to close down. After thirteen difficult years of prohibition the Stewarts concentrated on growing a pharmacy chain of stores. Finally, the distillery was able to once again open its doors and start producing whiskey and gin on a large scale—openly and legally.

Paul Stewart's new wife Angelique gave birth to four sons, all of whom went into the family business. Between them they took on running the distillery, managing the farm and growing the pharmaceutical factory. Andrew's father was the youngest son—he was born in 1933, the year prohibition ended. His father had invited Andrew onto the board of the pharmaceutical business and trained him and groomed him to take over the day to day running of the business.

When Andrew's grandfather died, the four sons took over the four aspects of the business: the farm, the distillery, the pharmaceutical factory and the growing chain of retail stores.

All four sons did well and became rich. By the time Andrew finished his studies—biological chemistry and pharmacology at Harvard Medical School—he was invited to sit on the company board and soon he was promoted to take on the Research and Development department. This suited Andrew well because he loved to travel and discover new drugs and recipes for cures. This was his way of contributing to society—helping find new remedies and new ways to relieve suffering and pain. He was known to joke that his grandfather had brought Jesus to the jungle, but he was also bringing the wisdom of the shamans to the western world.

Andrew was well travelled and had advisors and associates on all five continents. A tall man with ginger hair, freckles and glasses, he looked younger than his age and had an unassuming manner about him that made it easy for him to make friends and relate to people of different nationalities and beliefs. He was very good at languages, too, and was able to communicate effectively with people in South America, Asia, Africa and several European countries as well.

"What's up?" Andrew asked as soon as he was told by his secretary that Anthony was on the phone and as soon as the connection was made. Anthony sat back in his chair and put his feet on the desk.

"Have you ever heard of The Legrande Syndrome?" he asked.

"No," Andrew replied. "What's that? It seems there is a new syndrome popping up every day. You have probably heard of the "lazy legs syndrome"—that one was invented by our competitors. It doesn't really exist, but if you can convince enough people that it does, and sell them enough pills to cure it, you can make a cartload of money. And that is precisely what has happened."

"Well, that's largely thanks to the pharmaceutical companies," commented Anthony. "There certainly wasn't a lazy legs syndrome when I was growing up. But if you spend your life watching television and don't exercise enough, then eventually your legs will protest. I don't think taking a pill can help very much."

"Visor has come up with the latest syndrome, called "texting thumbs syndrome," said Andrew. It is real in that if you keep using your thumbs for texting, the thumb muscles will get tired. It is a little bit like shin splints—

only these are thumb splints. So we have developed a pill for thumb splints and I am sure it will make Visor a lot of money. But anyway, back to the reason for your call. Who invented this Legrande Syndrome?"

"Legrande discovered it; he didn't invent it," said Anthony, somewhat defensively. "I think this is a real syndrome. It's appearing in men all over the world. The numbers are still low, so we are not calling it an epidemic yet, but it is growing every day."

"So what is it?" Andrew was sounding impatient now.

"It's to do with the male sperm…" He was just about to continue when Andrew interjected. "Is there any other kind?" he asked.

"Of course there is," replied Anthony, now becoming irritated. "There is whale sperm, bull sperm and the sperm of thousands of other species. But that's beside the point. The point is that men are losing their ability to reproduce. Their sperm are losing their tales and are going nowhere. They seem to have no desire or inclination to travel up the woman's fallopian tubes."

Andrew tried to say something, but Anthony spoke again. "Before you say anything, yes, they tried to fertilize an egg *in vitro*, but that did not work either. The sperm just wriggle around, but do nothing productive at all."

"That's interesting," said Andrew. "Perhaps there will be fewer people in the world. That could be a good thing," he added as he started to think about the repercussions of such a state of affairs. Then he asked, "How do I come into the picture?"

"We need a cure," Anthony said. "We need more

research," he added. "You have the connections, the scientists working for you and the lab."

"All right, we'll have a look and see what we can do. Can you assure me that we will get an exclusive patent?"

"If you are the first to come up with a cure, then certainly, you can have exclusive rights," Anthony assured him. "But we can't wait indefinitely. Get your minions onto it and get back to me by the end of the month. I will send you all the material that I have on the subject."

"I will need sperm samples too," Andrew said.

"I'll arrange that, too," Anthony assured him.

# CHAPTER EIGHT
## The Search Is On

Four researchers were sent out immediately—one to China, one to India, one to Brazil, and one to Australia. Andrew knew that aboriginal medicine still held many secrets and effective cures that were as yet unknown to the western world. All the researchers had connections to local holy men, shamans, gurus and healers who knew the local herbs and remedies for all kinds of ailments and diseases. They travelled into the jungle, hiked up mountains and canoed down rivers in search of a cure.

Leon, the man who went to India stopped at an ashram where the famous healer and guru, Baba Indus had gathered hundreds of sannyasins (students) of all ages and origins from all over the world.

Every morning the guru held darshan, which was a period of interview with the guru, during which individuals could ask questions and receive answers. Guru Baba Indus sat cross-legged on a large gilded armchair on a dias, while the students sat cross-legged on mats in front of him.

Leon's parents were originally from India but he was born in the United States and only started visiting his parents' homeland for the past five years, since he started working as a researcher for Visor Pharmaceuticals. He did speak Hindi, but with an American accent that made local people smile indulgently at him.

"I want to petition for a healing," Leon addressed the guru when his turn came. "Not for me, but for someone back home."

Baba Indus listen attentively, as he always did, with his head slightly tilted to one side so he could hear the petition with his good ear, as he was deaf on the other ear. He had two women students whose duty it was to write down all petitions and the guru's responses. These were then printed in the local ashram newsletter, or, if necessary, his promises were followed up at the appropriate time.

"Meet me in the healing room after dinner," the guru said, and one of the young women sitting on a mat beside the guru's chair quickly wrote down the appointment.

Before Leon entered the healing room after the communal dinner served in the ashram dining room, he was informed by one of the helpers that healing, as well as time spent with the guru, was free, though priceless, but that donations were appreciated as they would allow the ashram to continue and grow. Leon had been given a small budget to be spent at his discretion in search of a cure for The Legrande Syndrome. He quickly offered what he thought to be a reasonable amount in the hope that he would be the one who could bring back an effective cure. His success would no doubt earn him a rise, and perhaps even a promotion.

When he entered the healing room he stopped for a moment at the door as his eyes adjusted to the subdued lighting. There were votive candles burning at several points in the room. To his right was a healing table surrounded by delicate mosquito nets. Next to the healing table was a lower table with bowls of water, crystals and flowers, as well as a statue of Ganesha, the Indian elephant-headed god.

The guru sat cross-legged on a dias at the other end of the room. Next to him on a mat was one of his helpers whom Leon had seen earlier in the day during dashran.

"How can I help you?" asked the guru in perfect English, pointing to a mat in front of him. Leon bowed to the master, then approached and sat on the mat, crossing his legs with his hands resting on his knees.

"I work for a pharmaceutical company," Leon said. "We are looking for a cure for a new phenomenon. It seems that there are men all over the world who are beginning to produce sperm without a tail. These sperm have no desire to perform their reproductive duty and impregnate an egg."

The guru smiled at this description. Leon had heard that the guru had a sense of humor and was known to often tell jokes during his lectures and deliberations, sometimes quite bawdy or even riské. He was a little bit surprised at his own boldness but was pleased to note that the guru seemed to appreciate his sense of humor. In the brief silence that followed Leon also noted how much easier it was to be courageous on behalf of other people. For all he knew, he might be undertaking a mission to save the world from depopulation and eventually demise. He had the thought that if he were petitioning for himself, he would be speaking in quite a different way.

The guru shook his head. "I have never heard of such a thing," he said. "The Lazy Sperm Syndrome," he added, clearly pleased with his new description of the predicament. "Have you tried fertility drugs?" he then asked. "You have so many," he added, shaking his head side to side.

"We did," Leon said. "They don't seem to make a difference."

The guru thought for a moment.

"No, I guess they wouldn't. It needs to be something that will target the sperm factory, where the sperm are made," the guru said, pointing to the relevant location.

Leon smiled with embarrassment. "How can we do that?" he asked.

"I don't know but I will mix you a potion that will target those little tadpoles before they are born. Perhaps that will help. I will bless the mixture with my special blessing which should give the relevant cells enough energy to fulfill their job properly."

The guru got up from his dias and walked toward the table with herbs and flowers. His helper also got up from her mat and followed. The guru opened a cabinet which was standing against the wall and extracted a small bottle. He then started picking up with his fingers small amounts of various powders and dropping them into the bottle. When he finished he shook the mixture up and

# The Search Is On

then murmured a few words over it that Leon could not understand. He then put a cork in the bottle and handed it to Leon.

"This should help," he said.

"Yes, but if it works, we will need to mass produce this mixture and gain approval from the FDA," said Leon. "I need the recipe."

"Well, I can give you the recipe. But only I can bless this medicine. Without the blessing it won't work."

"That might be so, but how can we present your blessing as a bona fide ingredient to government authorities who approve such things for general consumption?"

"You can't," replied Baba Indus. "But you can get approval for all the other ingredients."

"Yes, but you said yourself that it won't work without the blessing."

"But it will work if I bless it. They don't need to know that I had blessed it."

"We can't keep traveling to India every time we need to produce a new batch of the medicine," Leon said, and he could feel his frustration rising.

"Don't worry, relax." The guru placed a hand on Leon's shoulder. "I will bess your precious medicine from a distance."

"How will you know..." Leon started to speak, but the guru interrupted him. "I will know," he said and he then repeated once again, "I will know. Ranita here will give you the recipe," he added. Ranita tore a page out of the notebook she was holding. "Here you are," she said as she handed the page to Leon. "Take a spoonful dissolved in a glass of water first thing in the morning and last thing at night." Leon was about to protest that the medicine was not for him, but he thought better of it. He thanked Ranita and the guru and left the healing room.

Just as Leon was about to leave the ashram on the way to the airport, the guru's assistant approached him in the courtyard. "You do realize that the potion we gave you yesterday and the recipe are not for the man?" she asked.

"No," Leon replied, confused. "Who is it for?" he asked.

"It's for the man's partner," the assistant said with a smile. "It will help cure her anxiety caused by her inability to have children." With that the young woman turned around and left. Leon was more confused than ever. Once he regained his wits about him, he felt disappointed that now he was returning home with just some kind of calming medicine, rather than a cure for men who could only produce tailless sperm.

Angie was the young woman who travelled to Brazil. She had studied with a shaman before and had written a book about the medicines of the jungle. She believed that all of humanity's ailments and woes could be cured with the natural produce of the jungle. There seemed to be a remedy for everything within the natural worlds; it was just that western medicine was not ready to fully embrace herbalism and natural remedies to replace pills and chemical compounds that tended to have such dramatic side effects. Angie was not a doctor, but she had studied to be a nurse and and during her short career on a cancer

# The Search Is On

ward she had seen plenty of ill effects of the medicines the doctors had prescribed.

Angie was sensitive to human suffering and in fact she had lost her faith in a compassionate god as she witnessed her mother dying of lung cancer. She had decided to dedicate her life to helping people alleviate their pain; this is why she had decided to become a nurse. She also researched herbalism and studied Chinese medicine in her spare time. She never opened a practice, but she would help friends, neighbors and friends of friends when they fell ill and when prescription medicines did not seem to help. She was recruited to join Visor Pharmaceuticals to specifically conduct research into herbs that could alleviate pain. She had found that certain combinations of extracts from a tree bark, a root and a seed, as well as the poppy pollen, became the basis for a new pain killer. The way it worked, like all pain medicines, is that it reduced the pain stimuli issued from the brain, but did not totally negate the effect of the pain. So the person could still feel something, but it was no longer causing so much discomfort that the person could no longer function normally. In fact, the compound became popular on the black market because if taken by a person not in pain, it would produce a pleasant tingling sensation all over the body. It would make a person feel very much alive and in touch with all parts of their body. It could only legally be obtained by prescription, because it did contain small amounts of poppy pollen, which were inked to heroin.

Angie was sent to Brazil because she had been there before and because, being originally from Portugal, she spoke Portuguese. Apart from studying Chinese medicine, Angie had spent three months in the jungle, studying with a healer and a shaman, who became known as Joseph. He had shown her the healing properties of various trees, bushes and plants, many of which she could not find in her herbalism encyclopedia, a well-thumbed book she treasured and often consulted.

To arrive at the village where Joseph lived, Angie had to travel up the Amazon river. She recruited a guide and hired a boat with a crew of two young brothers who assured her that they knew Joseph and that they would take her to his village.

The journey only took two days, but traveling in the small boat took its toll, and by the time they arrived, Angie felt exhausted and bitten to bits by mosquitos. One of the brothers—the older of the two—had a cell phone, but there was no reception in the small village. So when it would come to the time of her return, Angie would need to find a local guide who could take her back to Manaus. Luckily for her, there were villages and settlements further up the river, and there were a couple of boats which would pass by the small settlement where Joseph lived once or twice a week. The difficulty was in knowing when the time for her return would be so that she could hail the boat to shore and travel back in safety. Angie knew the ways of these remote villages and trusted that all would be well. All she needed to do was listen to her instinct and stay as long as it would be necessary to procure a cure for this Legrande Syndrome.

When the boat pulled into the shore, Angie got out and said goodbye to her crew. From here she knew the way—a narrow path leading from the shore to the circle of huts around a central ring of compacted earth. This was the main meeting place and where all village rituals and activities were performed. As Angie approached the houses, she could hear chanting. She realized that there was a gathering in the middle of the village. People were dressed in colorful costumes; they were gathered around a raised dais. There, in the middle of the elevation, sat Joseph, and next to him there was an empty seat.

A small child, perhaps six or seven years old, came running toward Angie and without a word held out his hand to her. He had obviously been waiting for her. Angie took the boy's hand and they both proceeded to enter the circle of villagers. The child led her to the empty seat beside Joseph. Angie knew better than to ask questions, so she sat down and patiently waited while the villagers continued to chant.

Several minutes passed and Angie had a chance to look around and take in the scene: the huts, the jungle beyond the village, the gathered villagers, their bright costumes and face paintings, and Joseph, nodding his approval as they listened to the eerie melody. Then suddenly the chanting stopped and a silence fell over the gathering.

"I have been expecting you. I know why you are here," Joseph said, as he turned toward Angie. Angie smiled at him and waited for a further explanation. She had experienced the chief's clairvoyance abilities before.

"What you do not understand in your country," Joesph started to explain, "is that the human model is an experiment in process and is not finished or complete.

# The Tale of the Tailless Sperm

Despite all your wars and global epidemics, there are still too many people on Earth. As the next human iteration makes its appearance, the planet will not need so many of us to process the Earth energy.

"This new syndrome cannot be healed—it will continue to make its appearance as it helps control the number of human beings being born each year.

"Giving birth is not a right, but a privilege. It will become rare for women to bear children and the population of the world will diminish.

"However, the human will be subject to evolutionary change. The new man and woman will become clairvoyant, telepathic and sensitive to the universal energies that travel to this planet and make their appearance here.

"Today, for example, is the full moon. Did you feel its energy as you traveled here down the river? That magnet in the sky exerts its power and causes the tides to increase and wane. You are made mostly of water, so what do you think does the pull of the moon do to you? Then there is the influence of the sun and all the planets in our solar system—they all have their unique nature of energy which appears here and influences our behavior and moods. Astrologers know this and can predict when certain events are likely to happen in our lives."

Joseph fell silent for a moment and Angie saw this as an opportunity to ask a question. "When is this going to happen?" she said.

"It is happening now," Joseph responded. "Have you noticed that there are fewer births in the world these days?" Angie shook her head side to side for "no."

"It is happening. The planet does not need so many people. The recent pandemic took the lives of many older and vulnerable people. There is another, worse pandemic coming.

"There are people in the world who believe in justice and are dedicated to the truth. One of the jobs of the human is to process energy on behalf of the planet—energy that she cannot assimilate directly, and energy that she does not receive from organic life. But when the human processes low energy, that is energy that is no longer nutritious to the planet, she will find ways to diminish the pollution and the flow of toxicity."

There was another short pause while Joseph took a deep breath and continued. "Our rain forest is being destroyed and so are all natural habitats in the world. What has taken thousands of years to grow and develop can be burnt in a day. Most people are greedy and short-sighted, and until we change our ways and become caring stewards of our planet, she will retaliate for our wasteful behavior by taking away our ability to have children. Parenthood is not right, as many people think it is, but, as I said before, it is a privilege and an honor."

In that moment, Angie understood that there was no cure for The Legrande Syndrome and that planet Earth was making the decisions about who could and who could not have children.

Edgar's father was aborigine Australian and his mother was Irish. They met when his mother visited Australia and traveled to Uluru, the sacred rock near Alice Springs. His father was a guide who made his living by leading walking tours to inform visitors about the local fauna and flora, bush food and the Aboriginal dreamtime stories of the area. They met during one of these tours, and Edgar's father asked to meet his mother for a guided walk in nature the next day. During their walk they decided to form a team and work together to protect and learn about the local natural environment. His mother never went back to Ireland and Edgar was born two years later. His parents still lived near Alice Springs and Edgar was glad to be able to visit his parents at the cost of the pharmaceutical company he worked for.

Edgar knew a couple of tribal elders who lived near Alice Springs. They agreed to meet with him and invited him to the home of the older of the two.

"These are changing times," the older man said as soon as the three of them settled down with cups of tea. "The magnetic poles are shifting and the climate is changing, too."

"We are experiencing more earthquakes, floods, hurricanes, typhoons, and fires than ever before," the younger man added.

"Humans are in peril," the elder continued. "Evolution is on the move."

"What do you mean?" Edgar asked, surprised.

"There are too many people on the planet," the elder continued to explain. "It is a good thing that there will be fewer children born. Be pleased," he added. "This is a natural process. There is no drug, pill, injection, or potion that can cure this new phenomenon. Why do people in the West insist that they have the right to have children? It is a privilege, not a right. Not to mention that fact that many people who do have children are not qualified to be parents. There really should be a training and an exam before a couple is allowed to have children."

Edgar was quiet, thinking about what the elder had said. He did not have children himself and had not really given parenthood much thought.

"So you are not able to help cure this Legrande Syndrome?" he finally asked. The elder gave Edgar a penetrating, and what he interpreted as a sad and solemn, look.

"The world is changing, Edgar," he said, "and no one can stop the onset of evolution."

"But how is the inability to have children a symptom of evolution?" Edgar asked.

"You will see," the elder replied. "There is a new human appearing on planet Earth. He or she will not be influenced by the material worlds, but will be more attuned to energies—universal energies, planetary energies, the energies of relationships and the energies manufactured by how we think, feel and act."

Edgar could not imagine what such a human would look like, but he felt he had asked enough questions. He realized he had not received what he came to Australia for, but he received something far more precious in return: a hope for the future and the settlement that came with the realization that there was a deliberate plan behind the Legrande Syndrome.

Wu-Yin knew of a monastery in the mountains near Lijiang. It was only minimally affected by communist policies adopted by the government in Beijing. Wu-Yin had been there when he was still a student, and had even considered joining the monastery and becoming a Buddhist monk. But he was still young then and the senior monk at the monastery pointed out to him that he did not necessarily yet know what he really wanted to do with his life.

Thee were rumors in the monastery that Wu-Yin was a reincarnation of a monk who had died twenty years earlier, and they even showed Wu-Yin photographs of the monk, insisting that he looked just like the deceased Buddhist. Wo-Yin had to admit that there was a certain similarity, but as far as reincarnation was concerned, he was not at all convinced. Nevertheless, he was flattered to be acknowledged as one of the monks, even though it was in another lifetime.

"Come back in ten years, if you still want to become a Buddhist," the old monk had said. "In the meantime, finish your studies, and check out your options, and find out more about who you really are and what your spirit is advising you to do."

Wu-Yin was both upset with the monk's counsel, but also somewhat relieved. Deep down he knew the monk was right, and also there was the pretty little Mexican girl he had left behind at home while traveling to the country of his ancestors. He wanted to pursue a relationship with her and was not sure how long she would be able to wait for him, despite her protestations of love he was receiving almost daily over the phone.

Now, years later, he was traveling back to the same monastery he had visited in his youth, hoping that the monk who was in charge of the monastery would help

# The Search Is On

him find a cure for this new disease called The Legrande Syndrome.

It was not an easy trip, and Wu-Yin had to climb mountains and walk for several days to get to the monastery. Luckily, he was still fit and familiar with outdoor living and camping. He knew the area and the trail that led to the monastery.

On his arrival he was greeted by a young monk in orange robes, whom he had not met before. The monk led him to a room in the monastery where he was advised to wash and change into the red robes of a novice.

"Reverend Lin-Sou will see you in half an hour," the young monk said. "He will offer you a meal of rice and miso soup," he added with a smile, not sure whether Wu-Yin would consider such fare as a delicacy or not.

Wu-Yin thanked the monk who then departed and proceeded to prepare for the meeting.

Lin-Sou was waiting in his private quarters. He was sitting on a mat and the evening meal for two was displayed on a low table in front of him. He gestured to Wu-Yin to sit opposite him. Wu-Yin bowed in reverence and took his seat.

"I expected you," Lin-Sou said with a smile. "I know you are not here to join the monastery," he added. He then picked up his bowl of soup and drank from it. Wu-Yin followed his example and tasted the soup as well.

"Ah, very good," Lin-Sou said. "We are very grateful to have food to eat and a place to sleep this day." Wu-Yin nodded, expecting further revelations from the master.

"You have come from afar," the old monk continued.

"I know why you are here. They have sent you on a fool's mission."

Wu-Yin looked at the monk sitting opposite him in surprise. "How so?" he asked. He then realized his brief question might have appeared rude to the monk, so he added, "What do you mean?"

"The time has come for there to be fewer children and fewer people on Earth. You see, I am old, and my genetic is among the youngest on Earth and the least evolved, because evolution is in process and my genes have had less time to evolve before I was born than yours. The child who is born today as we sit on this mountain has the oldest genetic in the world and is the most evolved."

Wu-Yin was having trouble understanding. Lin-Sou continued. "You have Neanderthal genes in you and so do most people on Earth, with the exception of people born in Africa. You also have genes that have evolved over the centuries and that allow you to understand quantum mechanics, complicated equations, and the concept of time travel. They also help you achieve intricate solutions to modern problems.

"But most people refuse to develop their potential, or are simply ignorant of the fact that they have the responsibility to further their evolution. So instead of developing their spiritual abilities, like healing and clairvoyance, or mental travel, they want to amass material possessions and power.

"People in the West are mostly beguiled by ancient religious rituals which are no longer effective. We need to find new meditations and create new mandalas to invite the new angels and entities to help us regain our status as custodians and stewards of the natural worlds.

"The riches of the Earth are not there for us to exploit. We should be stewards of the Earth, and we should be considering how to enhance the world, not abuse and destroy it.

"So, in the meantime, there will be more natural disasters caused by climate change—fires, hurricanes, tsunamis, floods, tornadoes on the one hand—and fewer births on the other.

"But there will be those who perceive these changes and welcome them. There will be groups of people learning about energies, becoming healers and teaching others. Look for these people and you will find the precious few who will be able to have children in the future. They will be few and far between, but the human race will continue. That is the promise of the gods."

Lin-Sou smiled at Wu-Yin who sat in front of him silent and dumbfounded.

"Eat your rice or it will get cold," Lin-Sou said. "You need to keep your strength for the journey back."

# CHAPTER NINE
## Jenny's Dilemma

When Dick came back from work, he could see that Jenny had been crying. She was sitting at the kitchen table with a cup of tea in front of her and a tissue in hand.

"What is the matter, Jenny?" he asked, briefcase still in hand, as he wondered whether it was something he had, or perhaps had not, done.

"I don't understand," she sobbed as he pulled up a chair and sat down beside her, placing his briefcase on the floor.

"I don't understand," she repeated, "why we can't have children." Dick sighed a sigh of relief. Always ready with advice and an answer to the world's problems, Dick was quick to console his wife. "We can adopt a child," he said. There are a lot of children in the world who need a good home."

This did not seem to help, and Jenny wiped away some fresh tears from her eyes. "It's not so easy," she said. "And it is very expensive. Julie has adopted a girl from China, and Anne adopted a child from Russia. They both paid huge amounts of money for their charges. And then there is no guarantee who these children will grow up to be like. Why can't I have a child of my own, who will cary our DNA?" She looked at Dick as if he might have an answer to this dilemma.

"I really don't know," he sighed, feeling helpless and somewhat impatient at the same time. "I am sure Dr. McNeil can help us. We just have to be patient." Dick suddenly had a revelation. "Aha," he exclaimed, "that is why patients are called patients. They need to be patient because a cure takes as long as it takes and cannot be rushed!"

Jenny smiled through her tears, but her expression soon changed again. "I am tired of the pills, and the suggestion that my eggs need to be harvested, and your sperm not being up to much," she complained.

Dick was beginning to feel guilty. "It's not up to me," he said. "It seems to be a world-wide trend," he added.

Jenny looked surprised. "What do you mean?" she asked.

"There's an article in the paper," he responded. "The global sperm count is going down." Jenny again smiled through her tears at that turn of phrase. "So, you see, it's not just me," he added.

"In that case, we should deposit some of your precious disappearing sperm into a sperm bank," she said.

"That's a good idea," Dick said, thinking at the same time that it was not a good idea at all. If his sperm were not fulfilling their reproductive duty now, he could not see why saving them should make a difference. "It's a pandemic," he added, thinking that the scale of the phenomenon might interest Jenny. "They don't know what is causing it, but I wonder if this planet has decided she has had enough people. Maybe nine billion is just too many for her to look after."

Jenny smiled. "Do you really think she is intelligent?" she asked and then corrected herself. "I don't know why we are calling it a her," she said.

Dick was glad that the conversation was moving away from his sperm, and the possibility of freezing them, to a more global topic. "It makes sense to me," he said. "She supports us like a mother. And isn't she usually referred to as Gaia? That must be a feminine name. Of course," Dick added, "Gaia was the mother of the Titans and of all life. In Roman mythology she was Terra." He was glad his interest in mythology was paying off.

"Nevertheless, it is just a lump of rock in space," Jenny said. "And who were those Titans?" she asked after a moment of silence.

"They were the pre-Olympian gods and they got defeated in the war of the Titans. There were six masculine Titans and six feminine. The Greeks believed in equality between the genders," he explained. There his knowledge of Greek mythology ended, and Dick hoped Jenny would not ask any more questions.

"If the two of us have one child, or even two, we are not contributing to the population growth on Earth," Jenny said. She sounded like she was pleading now, hoping for his approval.

"But what if my sperm continues to be tailless? Would you accept a different sperm donor from the sperm bank?" he asked.

Jenny looked at him wide-eyed for a moment, as if she was trying to comprehend what he had just said. "I couldn't do that," she replied. "Who knows what someone else's child would be like," she added.

"They do give you a description of the donor," Dick explained. "They let you know his level of education, the color of his skin, eyes and hair," he added.

Jenny had made up her mind about sperm donations, and she was not about to change it now. "No, I want our baby," she said, in a whining kind of way.

"But maybe you can't have one," Dick tried to suggest, realizing that he was fighting a lost battle.

"I will keep trying," she said, now becoming adamant. Jenny could be stubborn and Dick knew it very well. He also knew there was no point arguing or reasoning with her when she was like this.

Then Jenny said something that surprised Dick. "Instead of buying sperm from a bank, why don't I just take a lover?" she asked. It sounded so ridiculous, they both burst out laughing at the contradiction.

Yet, in their laughter, Jenny could begin to feel her doubts about Dick and whether he was the right partner for her. After all, having a family was her most important goal in life.

# CHAPTER TEN

# A Global Issue

"This meeting is being held in strictest confidence," Bill Sedwick, the minister of Health and Welfare, said. "You may not speak to the press about it, or to your colleagues, or to anyone in your family or circle of friends. Did I cover everyone you know? You must not speak about this to anyone, and I repeat, anyone!" He looked around the room. "Do I make myself clear?" he asked. Everyone nodded and murmured, "Yes."

"Good," Bill Sedwick said. "Because if you do speak to the press or someone who might then speak to the press, you will cause widespread panic and depression."

Anthony was curious. What could be so important that it could cause panic? Surely Bill Sedwick was not referring to The Legrande Syndrome. As far as he knew, there were a few cases in different parts of the world, but it was neither a pandemic nor an epidemic. At least not yet, as far as Anthony was aware. Over the past weeks he had not seen any articles about the new syndrome, so he had assumed that it was no longer a threat and that it would not afffect the human race on a large scale. He was about to find out otherwise.

"We have a big problem—both in the country and abroad," Bill Sedwick said, making sure his words were having the desired effect of solemnity and menace as he looked around the room at the employees of the ministry. "It is affecting our ability to have children and can cause the population numbers to decline if we do not find a cure soon. He paused, took a sip of tea and continued. "Human sperm is on the decline—most of them have no tails and cannot impregnate an egg—whether *in vitro* or in the woman's fallopian tubes. The population of the world is declining. I have invited today four people who have been traveling the world in search of a cure, representing the largest pharmaceutical company in the country. They will share with us the results of their quest.

The people around the table looked around to see if they could spot the strangers in their midst.

"They are not here yet. They will be coming in a few minutes. I first wanted to meet with you privately to ensure that you would not speak of this new abnormality to anyone outside of this room. Please sign the agreement in front of you, in which you agree to keep everything you hear here today in secret."

For the next few minutes the people sitting around the table read the agreements in front of them, picked up the pens that were also placed in front of them and signed the printed documents. Bill Sedwick's secretary collected the pieces of paper and placed them in a pile in front of her boss.

"This is a serious situation," Bill Sedwick continued. "Let's hope the the representatives if the pharmaceutical company have some good news to impart." He then turned to his secretary and said, "You can call them now." Evelyn, who had been Bill Sedwick's secretary for two years, left the room.

Five people walked back into the boardroom: Bill Sedwick's secretary Evelyn, followed by Angie, Wu-Yin, Leon and Edgar. There were four empty seats at the end of the table, reserved for the young researchers.

"There are now thousands of people affected by this new Legrande Syndrome," Bill Sedwick started speaking as soon as the newcomers were seated at the table and the noise of shifting chairs had ceased. "The public don't know this yet, and we do not want to cause panic, but the world population is decreasing. These are the people dying every second," he said as he banged the table with his fist twice. "And these are how many are being born." He banged the table only once.

"If this trend continues, the birth rate will continue to diminish even further." The minister now addressed the researchers sitting at the end of the table. "You have gone around the world—to India, China, Australia and Brazil. What have you brought back and what can you tell us about the wisdom of the shamans, the gurus and the witch doctors, as far as this new pandemic is concerned?"

The room fell silent and the four travelers looked at each other, trying to determine who should speak first. Finally Leon raised his hand and began to speak. "I don't bring you a cure," he said, as he retrieved the phial of liquid from his pocket and placed it in front of him on the table. "What I am bringing is a medicine for the women who cannot have children. It has been tested and it does work as a calming agent. My company has taken the recipe and mass produced a pill." Leon pulled out of his pocket a small plastic box with a couple of pills in it. "It works, but it does not solve the problem we are facing," he added.

Angie spoke next, as she attempted to draw attention to the seriousness of the situation. "I have had a completely different experience," she said. "I visited the shaman Joseph in the Amazonian rain forest. He knew why I had come to his village but he could not help me. He said that the Earth was withdrawing her permission for people to have so many children." A few people around the table smiled at the idea that the Earth was intelligent and could decide the rate at which children were being born. "Joseph says the trend will continue." she added.

Wu-Yin spoke next. "It will continue because human evolution is not finished," he said. "But the good news is that the human race is not doomed, for there will be those who will be able to have children. These children will be evolved beyond the level of the indigo children, or even the crystal children. They will have the ability to see the energy in a person's aura and to instantly know who they are and what they are thinking, and whether their intentions are noble or wicked."

"The aborigines in Australia know this, too," Edgar added. "They know this is a time of evolutionary change. The poles of the planet are shifting and they are expecting the emergence of a new human model. But first we will have to put up with many natural disasters, before we get the point that we are influencing climate change and need to do something to rescue the Earth from extinction.

"It is clear that we will not grow a new arm or leg, but our brain has a lot of room for improvement. The few children who are born today and will continue to be born in the future will have the potential to develop many new abilities, like healing, telepathy and clairvoyance. They will be in touch with their spiritual guides and will be more concerned with their possibility to have a good reincarnation than amassing physical wealth in this life."

"This new message is coming through the religious orders in the world," Wu-Yin added. "At least in the Buddhist monastery in China it is known that the time for evolution is upon us. We should embrace The Legrande Syndrome and be grateful that it is helping bring the numbers of the human race to a sustainable level, and that we will be witnessing the evolution of the human race over the next decades, until only those who have the new faculties of clairvision and clairsentience."

# CHAPTER ELEVEN
## Epilog

Joe wrote the article. He had a friend who knew someone at the Ministry of Health and Welfare. This person, whom we shall call John, felt it was important for the American people to know what was happening in the world with the declining birth rate. He had secretly recorded the meeting at the ministry a few days before, and he sent Joe the recording, asking for Joe to keep his name out of the article.

Joe had listened to the recording several times. He tried to assimilate what he was hearing as best he could. He enlisted Maggie's help, because he always considered her to be more spiritually inclined than he was, and he believed that she would understand better the concept of evolution on the move.

He was amazed that the shaman in Brazil, the elder in Australia and the Buddhist monk in China all had similar perceptions about the change that was happening on Earth. It was both exciting and daunting at the same time, because he felt that all their western values and beliefs were being challenged at once. The changes he had already experienced in his lifetime, like climate change and the increase in natural disasters around the globe, all now seemed to be precursors to this final change, which was a new model of human appearing on Earth.

Joe thought of his next door neighbor's two-year-old daughter who was already fluent in two languages (her mother was French and spoke to her daughter in her native tongue), and also seemed to be very quick to grasp the meaning of even the most serious and in-depth conversations.

"Did you ever want to have more children?" Joe asked Maggie one evening, as he wondered whether The Legrande Syndrome had perhaps affected him in some way, as well as millions of men across the world. "We've been married for almost eighteen years, we have a beautiful teenager daughter, but it's funny that we have never really spoken about it," he added.

"I always thought that if I got pregnant again the natural way, we would have children, and if I didn't, which is what happened, we wouldn't," she replied with an endearing smile.

"When I write this article," Joe explained, "the world will know that the human race is on an evolutionary progression. Maybe we will start using a larger percentage of our brain," he mused. "I think I will entitle this article, *'Having Babies Is a Privilege, Not a Right.'*"

"This will probably be the most important article you have ever written," Maggie said. "It really changes all our previously cherished values."

"This whole situation proves that there is intelligence in the universe and that there are beings, or entities, or angels, or whatever you want to call them, that

# Epilog

are intervening in our life on our behalf."

"Or," suggested Maggie, "evolution is automatically set and a new human model appears when the time is right, just as the Neanderthal model disappeared all those years ago."

"I think the important thing to understand is that the human model is getting better."

"Though we have had some relapses when the model got worse, like during the Second World War."

"Yes, that is difficult to explain."

"However, more intelligence, healing abilities and clairvoyance are definitely hallmarks of an improved model."

"We certainly live in interesting times," Joe concluded, referring to the ancient Chinese curse.

Jean Legrande's article, which had come out several months earlier appeared and disappeared without much response from the readers of the magazine, much to Jean's chagrin. He continued his search and over the weeks he found many new patients who had the same syndrome.

He was writing a book and putting pen to paper, considering the repercussions on a global scale, if the world population would significantly decrease. It was an interesting intellectual exercise—imagining the world with fewer people.

When Joe's article came out, his phone did not stop ringing. Bill Sedwick was furious and the president was demanding who the source of the leak was and insisting that the information was a hoax. More and more men were being tested and more and more were learning that they, too, had the syndrome.

The message about human evolution was not easily acceptable in some circles. It seemed a far-fetched idea, but one that most people had an opinion about. But as time progressed, and fewer and fewer women could conceive, the reality began to dawn. The evolutionary change at first appeared in young children, who were now not only teaching grown-ups about computers and the Internet, but also about living in harmony with the Earth, revitalizing the soil and organic farming.

Children were becoming telepathic and clairvoyant, and it was no longer possible to lie to them about Santa Claus coming down the chimney or about the tooth fairy. They were quick to learn and achieving their degrees in record times. There were soon teenager graduates and professors in their twenties who began teaching their elders at colleges and universities.

It took time, but the planet started healing, as plastic trash was pulled out of the ocean, and fossil fuels were phased out over time.

We, who believe in the future,
shall not rest until it is here.

www.ingramcontent.com/pod-product-compliance
Lightning Source LLC
Chambersburg PA
CBHW031921240526
45464CB00021B/628